CW00393658

LAKESHORE LOOPS

First published in 2008 by
Liberties Press
Guinness Enterprise Centre | Taylor's Lane | Dublin 8
Tel: +353 (1) 415 1224
www.LibertiesPress.com | info@libertiespress.com

Distributed in the United States by
DuFour Editions | PO Box 7 | Chester Springs | Pennsylvania | 19425

and in Australia by
InBooks | 3 Narabang Way | Belrose NSW 2085

Trade enquiries to CMD Distribution
55A Spruce Avenue | Stillorgan Industrial Park
Blackrock | County Dublin
Tel: +353 (1) 294 2560 | Fax: +353 (1) 294 2564

Copyright © John Dunne, 2008

The author has asserted his moral rights.

ISBN: 978-1-905483-44-0
2 4 6 8 10 9 7 5 3 1

A CIP record for this title is available from the British Library

This publication has received support from the Heritage Council
under the 2008 Publications Grant Scheme.

Cover design by Siné Design
Internal design by Liberties Press
Printed by Colour Books | Baldoyle Industrial Estate | Dublin 13

This book is sold subject to the condition that it shall not, by way of trade or otherwise, be lent, resold, hired out or otherwise circulated, without the publisher's prior
consent, in any form other than that in which it is published and without a similar
condition including this condition being imposed on the subsequent publisher. No
part of this publication may be reproduced or transmitted in any form or by any
means, electronic or mechanical, including photocopying, recording or storage in
any information or retrieval system, without the prior permission of
the publisher in writing.

Lakeshore Loops

Exploring Ireland's Lakes

John Dunne

Contents

Foreword

I have to admit I am no cyclist. I can ride a bicycle but it's some time since I've done so. Despite this I've been really looking forward to Lakeshore Loops because I knew from John Dunne's previous publications that it was going to be about far more than just cycling.

I wasn't disappointed. I can imagine that people with a passion for cycling will find the clear descriptions of routes and facilities invaluable. Even more invaluable is the fair-minded way John praises what is good, and gently warns against what is more problematic. But these things only make up a small part of what is a quite considerable book.

Daniel Corkery's phrase 'the hidden Ireland' has been used and misused for nearly a hundred years now but this book amounts to a remarkable compendium of what it really means. This is a book about the path less travelled, about bits of Ireland – largely rural Ireland – that are off the beaten track. And it is a celebration of the heritage and culture of these places.

My own enthusiasm for bye-ways and backwaters developed from a love of travelling by boat. In the early days a lot of this travelling was by canoe, a quiet form of self-propulsion that has some similarities to cycling. And when I graduated to larger boats with engines, I always had some smaller craft in tow for exploring shallow places, accessing islands and allowing me to land on beaches and in little inlets.

The unspoken truth of the matter is that open water is rather boring. On the few occasions when I've been completely out of sight of land, I have found the experience less than inspiring. The real fascination is the shore – the place where the land meets the water. I have spent a lifetime exploring shores, mostly on our inland waters.

John, it seems to me, does the same thing. He approaches shores from the land rather than the water but they hold the same fascination for him as they do for me. And his approach is one of intense curiosity. He wants to know how the lake was formed, who lived in the ruined castle, why a place has such a strange name and what inspired the local poet. And his curiosity is not just confined to the historical past. He meets the people who still live in the area, and this, of course, means that he has a healthy interest in rural pubs.

We have a lot of lakes in this country and some of them are remarkably large. All the major ones, north and south, are included in this book. In the historical past, and indeed throughout prehistory, these lakes were important to people. They were valued, amongst other things, as transport resources, sources of fish and wildfowl and, of course, as water supplies. They also had a cultural significance in art and folklore, as John demonstrates.

In the modern age there has been a tendency for lakes to diminish in value. We have lost respect for them and they are often completely ignored. I think this is not only sad, it's rather dangerous. A loss of respect for the natural environment inevitably entails sowing the seeds of catastrophe.

This admirable book fights against that dangerous trend. It will open many people's eyes to a magical world they never knew existed.

DICK WARNER is a writer, broadcaster and environmentalist. His award-winning television series Waterways explored Ireland's inland waterways in twenty-four programmes and was followed by Voyage, a series about a circumnavigation of the island's coast in a sailing ketch. He has made many other radio and television programmes about the environment, heritage and wildlife. He has also written several books and is a regular contributor to magazines and newspapers, with a weekly column in the Irish Examiner. He lives with his family in rural County Kildare.

Acknowledgements

I once read somewhere that acknowledgements in books are boring and that nobody reads them anyway. I strongly disagree. Not only are they the chance to formally thank the many people who contribute to the creation of a book, but they also give a flavour of the journey through which a book passes, from the germ of an idea to the final printed outcome.

A book like this, which has a lot of factual content, requires many hours of library research, and I owe a deep debt of gratitude to the staff of the National Library and, at a more local level, of Stillorgan Library, for their assistance and patience. It would be impossible to thank in person the many authors whose printed words yielded much information that was useful to me in shaping the tours outlined in this book. I am truly indebted to them all but there are a few I would like to single out: Mary Rogers for her detailed account of Lough Erne, Gerard Madden for his meticulous description of Holy Island on Lough Derg, Jarlath Waldron for his fantastic account of the Maamtrasna Murders, and finally Frank Columb, whom I had the pleasure of meeting this year, for his inspirational book on Lough Gowna.

All the trail research for this book was carried out by bicycle, which, as the novelist Dorothy Sayers points out, is the most sociable form of transport. I am greatly indebted to the many people I met on my travels who enriched my journeys with snippets of information and local lore which have found their way into this book, as well as providing countless cups of tea and refreshments. Outstanding among those were Father Frank Fahey of Ballintubber Abbey, a truly inspirational human being, and Tom Quinn of Burriscarra, County Mayo, whose dreams of creating a special 'Doon Peninsula Experience' are constantly evolving.

The following organisations have also been of great assistance: the Northern Ireland Tourist Board, the Forest Service of Northern Ireland, the Environment and Heritage Service of Northern Ireland, the National Trust (Northern Ireland), Coillte, National Parks and Wildlife Service, ENFO (Department of the Environment, Heritage and Local Government), Inland Waterways Association of Ireland, Lough Neagh Discovery Centre and Emerald Star Line.

I am also very grateful to Dick Warner for contributing the foreword to this edition, and to the Heritage Council for their generous support towards the publication of this book.

Finally, I should like to thank Seán O'Keeffe, Peter O'Connell and the staff of Liberties Press for their faith, support and hard work in bringing this book to fruition.

Author's Note

While every effort has been made to ensure the accuracy of the information supplied in this book, no responsibility can be accepted for any damage or loss suffered as a result of error, omission or misinterpretation of this information. The author and publishers shall have no liability in respect of any loss or damage, however caused, arising from the use of this guide. This includes, but is not limited to, loss or damage resulting from missing signs, future changes in routes, accidents and lost or injured persons.

A number of the sites of interest mentioned in this book are situated on private property, and their inclusion may not be interpreted as an invitation to enter onto such property. Permission to enter must be sought from the owner in all such instances.

Introduction

The interlinking of cycling and water has been a passion of mine for many years. At first glance, these two things would appear to be strange bedfellows, but the discovery of towpaths alongside Ireland's waterways led me to a connection which I believe is a good marriage. My thirst for places to pursue off-road cycling had exhausted the trackways on the banks of Ireland's waterways. I had documented over 650 kilometres of cycling trails with one guiding rule: the trails had to remain on the banks of the waterways they tracked. As some of these waterways had coursed through or past lakes, the germ of an idea for further exploration presented itself: why not track the perimeters of our larger freshwater lakes? This was not going to be an off-road experience, but my concerns about competing for road space with cars and lorries were quickly allayed by the discovery that many of our best lakes are surrounded by small country lanes and farmers' trails which are ideal for cycling. With this knowledge, I set about trying to document lakeshore cycling trails.

When I first sat down to plan my research for this book, I made a list of the lakes that I thought would be suitable. I quickly noticed that an initial list of fifteen prospects soon started to extend. A quick glance at the map of Ireland shows a landscape that is pockmarked with lakes the length and breadth of the island. More intense study reveals that there are well over six thousand natural lakes in Ireland, varying in size from less than 1 km^2, to 381 km^2 in the case of Lough Neagh, the fifth-largest freshwater lake in Europe. More than 95 percent of these lakes are quite small, having a surface area of less than 1 km^2, but there are at least twenty-eight with a surface area of more than 7.5km^2; the focus of this book is these larger lakes.

There is a good geological rationale for our abundance of lakes. With the exception of Lough Neagh, Ireland's lakes were formed during the last Ice Age, which began between two million and one and a half million years ago. Moving ice sheets or glaciers up to a thousand feet thick chiselled the contours of the earth's surface, leaving behind deep troughs and gorges. When the ice melted due to rising temperatures, the troughs filled with water, to form our lakes. The water supply was augmented by the flows from rivers and streams.

Although the purpose of this book is not to investigate the creation of the lakes but to enjoy the benefits of their presence, it is important to note that the lakes are a major natural resource. In more modern times, some of our lakes have suffered neglect and destruction, either through pollution or as a result of drainage schemes that adversely altered their profile. In more recent times, however, there has been a growing awareness of the value of our lake systems and the importance of their connection with our waterways. Increased attention and resources are being devoted to redressing the ills that befell some of them in a less environmentally conscious time, and greater awareness is being developed of their value as an attraction for boating enthusiasts, anglers, cyclists, walkers and tourists in general.

There is evidence of human occupation along the shores of some of our more signficant lakes, such as Lough Neagh, dating back some nine thousand years. Communities of fishermen and hunters gravitated towards lakeshores because of the plentiful supply of fish such as salmon, trout and eel, and they used equipment, including wooden harpoons and nets of wickerwork and animal hair. Much later, followers of early Christianity established monastic settlements on the islands of some of the larger lakes, and up to about the eleventh century, aristocratic families built island dwellings known as crannógs. Lakes also proved popular among the new ascendancy class that emerged from the seventeenth century; some of the settlers were to build majestic homes on or close to lakeshores. Some of these, such as Belvedere House on Lough Ennell, County Westmeath, offer the modern visitor a chance to view the splendour in which the privileged few lived.

Ireland's lakes crop up here and there in poetry and prose, they obviously feature heavily in angling guides and scientific studies about water quality and natural history, and some travel writers, such as Sir

William Wilde, have written about specific lakes. Otherwise, though, very little has been written about them. Perhaps because of their abundance, we have learned to take them for granted.

Consequently, this book is not a mere clinical description of route details. It meanders through the history and heritage associated with our principal lakes and their hinterland and seeks to acquaint the reader with the legend and lore that has fuelled the imagination of lakeside communities. Hopefully, this guide will enlighten and entertain you as you make your journey along their shores, and help you to appreciate them fully. While the book was written with cyclists in mind, it will also be of interest to walkers or those who may prefer to tour by car. Indeed, anyone with an interest in legend and lore will, I hope, enjoy reading the book.

All the circuits outlined are designed as single-route loops where bicycles can be transported to the starting point. However, given the proximity of the lakes in a particular county or in neighbouring counties, the circuits could be undertaken as amalgamated tours, with readers choosing their own routes between the lakes. It is up to individuals or groups to make their own selections. I hope that the contents of this book will inspire in people a desire to visit as many lakes as they can.

As a final note, I should like to add that a lake is not just an amorphous body of water but a living thing whose story is constanty evolving. As Bill Bryson points out in his book A Short History of Nearly Everything, when you look at a lake, you are looking at a collection of molecules that have been there for about a decade on average. Lakes provide links to the past and continuously contribute to the creation of new pasts. They are also of vital importance to archaeologists: lake mud has perfectly preserved everything from domestic utensils to valuable artefacts. While they are no longer a central feature of people's lives, they are still a key aspect of our environment, and an invaluable leisure resource.

County Antrim

River Bann
Toome
Magherafelt
Moyola River
River Main
Randalstown
Antrim
Cranfield Church
Ballyronan
Ballinderry River
Ardmore Point
Crumlin
Ardboe Point
Ram's Isd
Glenavy
Castle Bay Centre & Quay
Scaddy Isd
Lagan Navigation
Coney Isd
Ardmore Point
Coalisland Canal
Kinnego Marina
River Blackwater
River Bann
Oxford Island Nature Reserve
Craigavon

Lough Neagh

(c. 190 kilometres)

LOCATION
To the east of Northern Ireland, within 32 km of Belfast city. Five of Northern Ireland's six counties touch its shores, the exception being County Fermanagh.

LENGTH
30 km/18.6 miles

WIDTH
12 km/7.5 miles at its widest point

AREA
381 km^2

PUBLIC ACCESS
Moving from south to north, the principal access areas are Oxford Island Discovery Centre, Kinnego Marina, Gawley's Gate quay, Antrim Lakeshore Park, Churchtown point, Toome, Ballyronan Marina, the Battery harbour at Moortown, Castle Bay Centre & Quay, Maghery Country Park, and Bayshore Picnic Area, Ardmore. There are many more informal access areas dotted around the lake.

MAP
Map 94 of the National Cycle Network Series available from the National Cycle Network Information Service (Tel. 0044 117 929 0888) or from Sustrans (Northern Ireland), McAvoy House, 17A Ormeau Avenue, Belfast BT2 8HD (Tel. 0044 28 9043 4569) (ISBN 978-1-901389-32-6) Also Sheets 14, 19 and 20 of the Discoverer series of maps available from the Ordnance Survey of Northern Ireland.

Lough Neagh lies 15 metres above sea level and is the largest freshwater body of water in the British Isles and the fifth-largest in Europe. Broadly rectangular in shape, it occupies an area of over 380 km² and has a shoreline extending to over 125 km. It is widely acknowledged that this shoreline was one of the earliest inland areas of habitation in Ireland.

Early origins

The lake's name is an anglicised variation of the Irish, Loch nEochaidh, which translates as 'the Lake of the God Eochaidh'. Eochaidh (pronounced 'Yeo-hee') was a horse god and a legendary king of Munster. There are many stories told in Celtic folklore and myth surrounding the origins of Lough Neagh. Among these is the legend that the lake was formed from the urine of a great magical horse on which Eochaidh had abducted his father's wife, Eibhliu. Eochaidh had been warned not to allow the horse to urinate anywhere while making their escape. Unfortunately, when they eventually came to a stop to settle in Ulster, the horse urinated, causing a spring to rise up on the spot. Eochaidh covered up the spring but a woman later drew water from the spring and omitted to replace the cover. As a result, the waters gushed up to form a lake, drowning all of Eochaidh's tribe, with the exception of his daughter, Li Ban, who later became a mermaid and spent the rest of her days wandering in the waters of the lake. Another tale with connections to Eochaidh is that in the first century AD he drowned at this location in a well whose waters suddenly burst forth and formed the lake. Perhaps the best-known legend relating to the lake is linked to that great Irish mythological warrior Fionn Mac Cumhaill. Fionn is said to have created the lake when he ripped up a great swathe of earth to throw at a Scottish warrior who was making his escape via the Giant's Causeway. His shot landed in the Irish Sea to form the Isle of Man. The scientific explanation is that the lake formed about 40 million years ago from a depression created in the Earth's surface by the collapse of basalt rock in an area that was made fragile during the volcanic era twenty million years earlier.

The lake gets its water supply from eight significant rivers, which empty their flow at various points around its shore. Only one river, the Lower Bann to the north, provides an outlet to the sea. It is relatively shallow: the deepest point measures only 30 metres, and the average depth is no more than 9 metres. This is some 3.6 metres lower than it was prior to 1847, when the first of several flood-relief and drainage schemes was started. In the mid 1960s, the unionist politician Captain Terence O'Neill suggested that the lake should be drained and filled in so as to create a new county called Neagh. Fortunately, his proposals did not see the light of day.

In the days when inland waterways were at their commercial peak, Lough Neagh was a hub of activity, with five canals connected to its shores. Today, its importance lies in the fact that it is the primary water source for one third of Northern Ireland's population. It also supports the largest eel fishery in Europe and a burgeoning leisure and tourism industry.

Ownership

As with any other property, ownership of our lakes is vested in a variety of private individuals and state agencies. In the case of Lough Neagh, the present owner is the Shaftesbury Estate, which receives income from the sand-extraction activities that take place on the lake. The estate also used to receive payments from the eel fisheries along the northern shore, but these payments stopped after many years of resentment on the part of the fishermen. The estate derives no income from the provision of drinking water.

It was Charles I who more than three hundred years ago granted the bed of the lake and fishing and hunting rights on and around it to Arthur Chichester, Lord Donegall. The title subsequently passed down to the Shaftesburys. The Shaftesbury family name is Ashley-Cooper and the family seat is a nine-thousand-acre estate at Winborne St Giles in England. The First Earl of Shaftesbury was an adviser to Oliver Cromwell and over the years a number of the earls have been prominent for a variety of reasons, not always positive ones. On the plus side the Seventh Earl campaigned vociferously against child labour and is commemorated by the statue of Eros at the intersection of Piccadilly Circus and Shaftesbury Avenue in London. However, it is said that the family is the subject of a curse. One ancestor collapsed and died while in a fight at Eton, and the Eighth Earl committed suicide by shooting himself. In more recent times, the Tenth Earl, a flamboyant and extrovert individual, was found murdered in the French Alps in May 2005. He had been missing for several months; it is thought that he was murdered at an apartment overlooking Cannes bay. His third wife and her brother were placed under investigation for his murder. His eldest son, Anthony, a London banker, succeeded to the title at the age of twenty-seven but died of a heart attack at his brother's apartment in New York a short time afterwards. The younger brother, Nicholas, a Manhattan-based music promoter and DJ, then became the Twelfth Earl.

The circuit

From a cycling perspective, Lough Neagh is the best documented of all the lakes on the island of Ireland. It has been included as part of the United Kingdom's National Cycle Network and is actively promoted by Sustrans under the designation 'Loughshore Trail'. An official route map and guide is available; this covers a circuit of 206 km around both Lough Neagh and Lough Beg, the small lake that lies north of Toome at the north-eastern extremity of Lough Neagh. The circuit is fully signed throughout and enjoys the benefit of a generally level profile, with the highest point, at Gortigal, only 100 m above sea level. The route outlined in this book follows reasonably closely that recommended by Sustrans and, given the comprehensive signage covering the route, this section will largely eschew route description – unlike other parts of this book. Instead, the emphasis will be on helping you get the best out of your journey by providing a flavour of the history and heritage of the lake's hinterland and the myth and folklore passed down through the communities that have dwelt close to the lakeshore.

Though the circuit is hailed in cycling literature and promotional material as being a highly attractive cycle route, my feelings about it are mixed. Perhaps my judgement has been tainted by the superior attractions of the circuits around many of the other lakes included in this book. Although the route has a huge amount to offer, there are a number of issues which take from the route as a pleasurable cycling experience. There is quite intense farming activity around the lake, and this leads to a number of drawbacks and dangers. The small country lanes used by the official route are in certain areas also used by farm machinery and livestock. This applies particularly on the eastern shoreline. The road surfaces in these areas are pitted with potholes and covered in mud and manure to such an extent that the metalled surface is not visible. In addition, the insects and flies that are a familiar minor irritant for any cyclist embarking on a countryside tour seem to abound in the vicinity of Lough Neagh. My advice is to keep your mouth closed at all times! Another gripe is the necessity to leave the lakeshore to embark on a lengthy diversion through generally uninteresting roads to Portadown in order to find the nearest crossing point over the River Bann. The detour is a lengthy 25 km. A footbridge at the estuary near the small village of Charlestown, similar to the one further north over the River Blackwater, is planned, construction is long overdue. There used to be a ferry across the river, but this has not operated for some time.

Oxford Island Discovery Centre

Although the circuit can be started at any point around the lake, I find that the most convenient set-off point is the Oxford Island Discovery Centre, located in County Armagh on the south-eastern corner of the lake. This is located on a small peninsula that was formed when the lake level dropped as a result of various flood-relief and drainage schemes undertaken in the nineteenth and twentieth centuries. The centre houses a natural-history exhibition and has information about nature reserves located around the lakeshore. There is also a coffee shop and ample car-parking facilities. Opting for a clockwise circuit, a choice faces you immediately you leave the entrance to Oxford Island. The alternatives are either to head straight across the M1 towards the Craigavon Lakes & City Park or to turn sharply right at the entrance on to a lane that leads to the Derrymacash Road. The second alternative allows you to miss out the Craigavon Lakes and to link up with the signed trail as it approaches the lakeshore west of Oxford Island. The first sighting of the lake comes after you turn off the Derrymacash Road to head for Ardmore point, and for a short while you remain on the lakeshore as you travel through the townland of Ardmore. Once you have returned to the Derrymacash Road, a looped extension allows you to visit the small but pretty village of Charlestown, where the road comes to a dead end. A small road called Lough Lane leads down to the estuary of the River Bann, and you can look forlornly across to the other bank with a slight sense of annoyance that, for want of a footbridge or ferry, you have to embark on a 25 km detour to reach the other side.

Portadown

Following the route of the River Bann southwards, you head back towards Craigavon, where the trail takes you through a shopping centre before you make your way to Portadown for the crossing of the river. The route through Portadown takes you past Craigavon Senior High School and the Upper Bann Institute, where you will come across one of a number of Millennium Mileposts that are scattered throughout this route. This one indicates a distance of 1.25 miles (2 km) to Portadown and 2.5 miles (4 km) to Craigavon Lakes. You avoid the worst of the Portadown traffic by availing of the cycle lanes through Eden Villa Park before crossing the river.

Portadown has been an important crossing point going back to early history. Its name is derived from an anglicised version of the Irish phrase Port na Dún, which means 'landing place of the fort'. This refers to a 'rath' or circular stronghold that overlooked the River Bann at this location. The opening in 1742 of the Newry Canal, the first commercially operated canal to be built in Ireland, and its proximity to Lough Neagh, meant that Portadown gained a pivotal trading position, and it soon became an important inland canal port trading goods via not only the Newry Canal but also the other man-made inland waterways such as the Lagan, Ulster and Coalisland canals. It remains a busy town, and you may experience a little difficulty finding the correct way out of the town, as some of the signposts are obscured from view. The exit from the town leads you through areas and streets that have become familiar names the world over and are a stark reminder of the recent Troubles. The route takes you past Drumcree Church and the Garvaghy Road before setting off into the countryside for the return journey to the lakeshore near Maghery.

Coney Island

A few hundred metres off the shore at Maghery lies Coney Island, which was once linked to the shore by St Patrick's Road. There are several stories as to how the island got its name. In one tale, dating back to the fourteenth century, Coney was the name of an old woman who lived nearby and who had a reputation as a healer. On one occasion, she was requested to help an Irish chieftain called Dermot O'Neill who had fallen ill on his way to do battle against the O'Connor clan. (The O'Connors had a castle on the shores of Lough Neagh; you can still see the ruins nearby at Milltown.) She refused to travel to his aid and eventually the O'Neills brought Dermot to her. Despite her efforts, Dermot died and the O'Neill clan suspected that she was a supporter of the O'Connors – a suspicion heightened by the fact that, for her own safety, Coney was installed by the O'Connors on the island that now bears her name. A less fanciful explanation lies in the fact that there are islands of the same name on other lakes in Ireland, such as on Upper Lough Erne (now called Inisrath); the name is thought to be a derivation of the Irish word for rabbit, 'coinín'. Boat trips to the densely wooded island are available locally, and it is worth a visit. Among the interesting sights are a thirteenth-century Anglo-Norman motte, St Patrick's Stone, where St Patrick is said to have rested while on the island, and O'Neill's Tower, which Shane O'Neill, chieftain of the O'Neill clan in the sixteenth century, used as a look-out post.

The Ulster Canal

Skirting past Maghery Country Park and the short Maghery Canal, you arrive at a footbridge, opened in the spring of 2002, that allows you to cross the River Blackwater. This bridge, known as the Ferrybridge, replaced a ferry facility and is an integral part of the official cycle route, with a Millennium Milepost marking its southern access. The River Blackwater was used to link the Ulster Canal to Lough Neagh. This canal provided a waterway link between Lough Erne and Lough Neagh and extended for 93 km, passing through the counties of Armagh, Monaghan and Fermanagh. The River Blackwater section accounted for 13 km of its course. The canal was opened to commercial traffic in 1842 and was eventually closed in 1929; since then, much of its infrastructure has fallen into disrepair or disappeared. There are proposals to reopen it, and perhaps when this has been achieved there will be the opportunity to take a cycle tour on its towpath through the heart of Ulster.

The Coalisland Canal

Not too far upriver from the Ferrybridge, another canal used the River Blackwater as a link to Lough Neagh. The Coalisland Canal, often referred to as the Tyrone Navigation, was built between 1733 and 1755 and was used to transport coal from the coalfields of east Tyrone. The canal joined the town of Coalisland to the River Blackwater and was 7 km long, with seven locks. This canal was later extended a further 4 km, to bring the navigation right into the heart of the Drumglass Collieries. This section is known as Ducart's Canal, after the engineer, Daviso de Arcort, who designed it. In constructing the extension, De Arcort introduced a new engineering technique that had not been used in Great Britain or Ireland up to that time. Instead of altering the water level by using locks, he used three inclined planes, also known as dry wherries or hurries – dry slopes where the coal would be offloaded into barrows and wheeled down to a waiting barge at the next, lower stretch of the canal. The new system did not work well, however. The remains of two of the inclined planes can still be seen today at Farlough and Drumreagh. Navigation on the Coalisland Canal ceased in 1946 and the canal was officially abandoned in 1954.

Lough Neagh Wetlands

Dominated by the estuaries of the Rivers Bann and Blackwater, this part of the Lough Neagh shoreline is known as the Lough Neagh Wetlands. It is an area of great importance for wildfowl and has a rich flora and fauna that attracts scientific interest. Looking around, you will observe a combination of fen, bog, woodland, small lakes, and low hills known as drumlins. On a windy day, it is quite exposed and cycling can be difficult. On the northern fringes of the wetlands is the townland of Brocagh, a derivation of the Irish word 'brocaigh', meaning 'ridge of the badger'. For some reason, the official trail heads inland over rising ground instead of following the shoreline, which is the superior and gentler option. At the crossroads at Brocagh, turn right, following the directions for Mountjoy Castle. This leads to the Castle Bay Centre & Quay, which was opened in 2001, and where there are fine views across the lake to the Antrim Hills. You will also be able to see the curiously named Duckingstool point, which lies about 750 m to the south of Castle bay. The name has its origins in a tradition that involved holding people who were suspected of witchcraft under water for a period. The end result was the same for anyone subjected to this torture: if they survived, their culpability was confirmed and they were put to death for their crime; if they perished, they were deemed to be innocent.

Ardboe Abbey and High Cross

You will not have great views of the lake for the next 7 km, as the route meanders along country lanes past rich farmland until you connect once again with the official trail and journey towards Ardboe point. Ardboe Abbey dates from the tenth century, when it was founded by St Colman. According to local folklore, the saint built the church using mortar mixed with the milk of a magic cow. Nearby, at the entrance to the graveyard, is a four-metres-tall high cross, which is badly weathered. The cross is recognised as one of the finest high crosses in Ireland, depicting biblical scenes in a series of panels carved on all sides. Unfortunately, part of the ring is missing and the capstone is badly eroded.

Ballyronan

Continuing past Battery harbour, the journey to Ballyronan keeps you in reasonably close contact with the lake. Not too far past the bridge over the Ballinderry river, which lies on the county boundary between Tyrone and Derry, you should once more leave the official trail and approach Ballyronan by the road that follows the shoreline. Today, Ballyronan's quiet demeanour belies its extensive history. The surrounding area has yielded evidence of some of the earliest human occupation along the shores of Lough Neagh, dating as far back as 7,000 BC. In more recent times, it was one of the last areas under the control of the Irish chieftains, before it was surrendered in the early years of the seventeenth century and succumbed to plantation by English and Scottish settlers. It developed into an important inland port around which industrial enterprises such as ironworks and tanneries thrived. In the nineteenth century, a passenger service operated from the port on a steamboat called Lady of the Lake, which was owned by the Gaussen family, originally a refugee Huguenot family from the south of France who came to Ireland in the early 1700s. The passenger service prospered while Lough Neagh had canal linkages. However, the arrival of rail transport, the closure of the canals and the lowering of the water levels in the lake caused by the drainage schemes of the 1840s heralded the demise of Ballyronan as an important inland port. Today, Ballyronan has rekindled its lakeside connections, with the construction of an impressive modern marina and visitor centre providing berths for fifty boats and two slipways.

Toome

The road to Toome is extremely busy and you will have to be especially careful near the sand-extraction plant that lies between the road and the lake on the outskirts of the town, as you will encounter quite a few lorries. Sand extraction from the bed of Lough Neagh is an important industry in this region. Another important industry, for which Toome is perhaps more famous, is the eel fisheries. The town has the largest eel fishery in Europe and is the headquarters of the Lough Neagh Eel Fishermen's Co-operative Society, which controls the catching and marketing of eels. Each year, it exports 500 tonnes of brown eels and 150 tonnes of silver eels. There are over one hundred boats licensed to fish for eels in the lake.

It is estimated that there are as many as 200 million eels in Lough Neagh. The type found in Irish waters is the European brown eel (Anguilla anguilla); each eel traces its origins not to the cool murky waters of an Irish lake or river but to the warm blue seas of the Carribbean, specifically the Sargasso Sea off the coast of Florida. The young eels, called elvers, arrive in Irish waters after spending one year drifting across the Atlantic on the current of the Gulf Stream, during which time they grow from 7 to 70 mm in length. They remain in Irish waters for between eight and as many as fifty years, having grown to a length of nearly 1 m, before making the 6,500 km return journey to the Sargasso Sea, where they breed once before dying. As they mature, they change from being brown eels to silver-black eels. The long life span of eels will surprise many. One female caught in County Mayo is thought to have been fifty-seven years old: the eel's age was measured by counting growth rings in a small bone in its skull.

Toome occupies a pivotal position, being the link point between Lough Neagh and the river that yields it access to the sea over a northerly course of 61 km. St Patrick is said to have built one of his many churches nearby in the fifth century. In the ninth century, the Vikings would have passed through Toome to gain access to the lake.

From Toome, you have the option to cycle northwards alongside the shore of the small lake called Lough Beg (which is more like a broadening of the Lower Bann) and continue tracking the river to Portglenone. This extension will add 38 km to your journey. The official route has the disadvantage that you cover the same ground twice, but it is possible to take a slightly longer course by passing from

Toome to Portglenone via Bellaghy on the western side of Lough Beg, and returning to Toome on the official eastern route. This will increase the Lough Beg extension to a distance of over 41 km.

Churchtown point and Cranfield Church

The official Loughshore Trail continues along a track adjacent to the Toome Navigation canal, which is part of a woodland park opened in 2002, and emerges on to country lanes that lead to Churchtown point. Here are the ruins of the seventeenth-century Cranfield Church, which was built on the site of an earlier thirteenth-century church. The name is derived from the Irish Creamh Choill, which translates as 'wood of the wild garlic'. It is said that St Olcan of Armoy, a contemporary of St Patrick, is buried within the church grounds in soil that was brought from Rome. Also on the site is a holy well believed to have been blessed by St Olcan; it is said that amber pebbles to be found at the base of the well have great healing powers. The pebbles are in fact crystals of gypsum. Up to the early nineteenth century, pilgrims used to visit the well in search of cures. They followed a ritual that involved doing seven separate circuits of the church and the well while reciting prayers and then bathing in water from the well. Apart from providing health cures, the well's pebbles were believed to protect women during childbirth, men from drowning, and homes from fire and burglary. It is a wonder that there are any stones left at the well! A tree close to St Olcan's well is festooned with rags – something that is found in many trees close to holy wells around Ireland. This follows a tradition founded in the belief that rags withering in the branches of the tree will draw infection away from the body if the infected part has first been bathed using a rag soaked in the blessed water of the well.

Randalstown

Travelling inland, you come to Randalstown, which is dominated by a nineteenth-century viaduct that spans the River Main. The town was originally called An Dún Mór, which translates as 'the great fort'; it was later known as 'Iron Mills' following the development of an ironworks mill in the town. Its modern name dates back to the middle of the seventeenth century, when Lady Rose Ann O'Neill renamed it in honour of her husband, Randal McDonnell, the Second Earl and First

Marquis of Antrim. In the nineteenth century, it became famous for its linen: at its peak, Randalstown's internationally renowned linen factory employed over a thousand workers. The factory closed down in the early 1980s.

You get a fine view of the town and the 174-hectare nature reserve adjoining it (known as Randalstown Forest) from the top of the viaduct, which has been converted into a walkway and cycle path. You will also be able to see across to the estate that surrounds Shane's Castle, the family seat of the O'Neills of Clandeboye, who secured title to the castle and its demesne in 1607 following the Flight of the Earls, when the remaining Irish chieftains fled Ireland for Continental Europe. The original castle was destroyed by fire in 1816 but the O'Neill family continue to live within the grounds of the demesne. The castle lays claim to its own banshee, a female spirit that is related to the fairy people and is renowned for its chilling cries. A local tale links the banshee to an incident when one of the O'Neill ancestors rescued a cow from a tree that was considered sacred by the fairy people. In doing so, he broke a number of branches of the tree. On his return home, he learned that, in revenge, his daughter Kathleen had been spirited away by the 'little people' to their home at the bottom of the lake. Her plaintive cries are said to be heard whenever an O'Neill family member is in danger.

Antrim Castle and the Clotworthy dynasty

My preference for the journey from Randalstown to Antrim is to use the wide straight road that runs along the demesne of Shane's Castle, where a cycle lane has been partially installed on the footpath on the south side of the road. The official trail opts for a more northerly route before linking up with the main road at Milltown. As you approach Antrim, you have to take care not to miss a turn immediately past Massereene Barracks, which leads you off the Randalstown Road past the Clotworthy Arts Centre and Antrim Castle Gardens.

These places are worth further exploration. Clotworthy is the family name of the Viscount Massereene, whose ancestry can be traced back to an Elizabethan officer, Captain Hugh Clotworthy, who came to Ireland in the late sixteenth century. In the early years of the seventeenth century, he was appointed High Sheriff of County Antrim. He was knighted by James I in 1617 and was subsequently granted lands around an old Norman motte that he had earlier fortified. The arts centre is located in what was originally the nineteenth-century carriage house and stables of Antrim Castle; it later became the Massereene family residence after a fire destroyed Antrim Castle in 1922. The building now houses galleries, small theatres and meeting rooms.

The remains of Antrim Castle can be seen in the magnificent gardens that were created along the banks of the Sixmilewater river from the seventeenth century onwards. Construction of the castle was started in 1610 by Sir Hugh Clotworthy, who died in 1630, and was continued by his son, Sir John Clotworthy, who was created the First Viscount Massereene by Charles II for his loyalty to the royalist cause during the Civil War period: he had been imprisoned for three years by the Parliamentarians. Among the other titles conferred on him was that of Baron of Lough Neagh. The castle was remodelled in 1813 by the Eighth Viscount, Chichester Clotworthy Skeffington. On 28 October 1922, the castle and its valuable contents were destroyed by a fire which took hold during a grand ball. As the fire raged, a young servant girl, Ethel Gilligan, was trapped in an upstairs room. She was rescued but had suffered severe burns; she died of her injuries on the lawn in front of the blazing building. It is said locally that a figure can be seen wandering the castle gardens dressed in white; this ghost is known as the White Lady. The ruins of the castle were demolished in 1970.

The gardens around Antrim Castle are one of only two surviving examples of Anglo-Dutch water gardens on the island of Ireland. (The other is at Kilruddery House on the outskirts of Bray, County Wicklow.) Among the many interesting features around the demesne are a parterre and grove enclosed within high lime and yew hedges, tree-lined canals, a twelfth-century Norman motte, a burial ground and a round pond.

An interesting tale is told about the wife of the founder of the Clotworthy dynasty. Sir Hugh had married Marian Langford, the daughter of Sir Roger Longford, an officer of the Lord Deputy, Sir John Chichester. Lady Marian fainted when she was confronted by a wolf while strolling through the woods near the castle. When she came to, she was surprised to find the wolf dead and an Irish wolfhound lying beside her. The dog followed her home and was taken into the household. One day, the dog disappeared – in much the same manner as it had mysteriously appeared. Some time later, the occupants of the castle were awakened one night by the howling of a wolfhound. When they lit torches, they discovered that enemy forces were gathering below the castle for an attack. At dawn, they noticed a stone figure of the dog standing on the castle's highest turret. The legend is commemorated by a statue carved in 1612.

Returning to the circuit, the trail leads past Antrim Forum. You will need to turn right here to make your way to Antrim's Loughshore Park, where you will find a welcoming café. Among the sights you will see at the viewing area are a derelict structure close to the shore. This is the remains of a torpedo platform that was erected by the Admiralty in 1942 as a test facility in preparation for the Allied invasion of Europe and North Africa. Today it is a breeding ground for common terns, who build their nests there every May – but will abandon them if disturbed. In deference to their needs, there is no public access to this structure.

Greenmount College walled garden

Leaving Loughshore Park, you get the chance to go off-road for a short while, availing of a 1.5 km trail (through Rea's Wood) which keeps close to the shore. Moving inland, you emerge near Greenmount College, renowned for its walled garden, which is open to visitors. The garden was built in 1801 and was originally used to supply produce to the residents of the Manor House and the estate workers. It was restored in the late 1990s; it now affords the horticulture students of Greenmount College the opportunity to gain hands-on experience while also offering visitors a chance to see how a typical walled garden attached to a great country house would have appeared.

Ram's Island

The official trail is well signposted and leads you back for an extended journey along the lakeshore, passing by Ardmore point. It will be recalled that you visited another Ardmore point on the southern shore in the early stages of the circuit. The sound of aircraft becomes a familiar noise: Belfast International Airport lies very close to the east. Staying with the aircraft theme, there is a former US airbase located close to the lakeshore at Lennymore bay; the airbase is now home to the Ulster Aviation Heritage Centre. The centre, which is open on Saturday afternoons from March until November, exhibits several old aircraft and related memorabilia. Sited in the middle of Lennymore bay is Ram's Island, the largest of the few islands to be found on Lough Neagh. The island is shrouded in tall beech trees, and these obscure the view of a number of ruins. A monastic settlement was established on the island in the tenth century; it was destroyed in 1112. Amongst the ruins of this settlement are the fragile remains of a round tower stretching up to 13 m in height. Also hidden among the trees on the island are the ruins of an O'Neill summer house.

Lagan Navigation

Pushing ahead southwards, there are excellent views across the lake as the trail continues very close to the lakeshore. Near the townland of Aghagallon, you get the opportunity to travel for a short while on what used to be the towpath of a section of the Lagan Navigation. This was one of five canal linkages that were developed along the shores of Lough Neagh during the 'Canal Mania' era of the eighteenth and nineteenth centuries. The Lagan Navigation was started in 1756 but a shortage of funds and problems with flooding caused delays, and it was not completed until 1794. The original plan was to canalise the River Lagan as much as possible and to use independent collateral cuts where awkward bends were encountered. This was achieved as far as Sprucefield near Lisburn, but from there on the designers eventually opted for the superior choice of a canal channel independent of the river. The canal was never a spectacular commercial success, although it did survive intact until it was eventually abandoned in the 1950s. Unfortunately for those who harboured ambitions to restore it as a navigable waterway, more than one-third of its length later became part of a less sedate transport route – the M1 motorway, constructed in the 1960s. There are plans to increase the use of the waterway, as a leisure-cruising facility, by cutting a new replacement section along a different route between Lisburn and Moira. In the meantime, attempts are being made to have the towpath cleared between Moira and the junction with Lough Neagh at Ellis' Gut, to create a shared-use pathway and incorporate it into the National Cycle Network's Northern Ireland routes. Some of this work has already been completed, offering those enjoying the circuit around Lough Neagh the chance to sample what the route will be like when completed.

It is possible to view the actual junction with Lough Neagh if one wishes. For the last kilometre of the canal's journey to the lake, the towpath from Annaghdroghal Bridge has merged with the adjacent fields, but the route can still be tracked. There are several fences to be surmounted and the ground can at times be marshy. As you approach the lakeshore, you will notice that the canal does not immediately cut straight across to the lake at the earliest point of potential contact. Instead, it runs alongside the lakeshore for about another half a kilometre to Ellis' Gut, apparently because the lake is too shallow at the earlier point. There are ruins of one house on the far side of the canal along this stretch, and two other dwellings have been extended and modernised. In two of these houses lived the men who operated the tugboats that brought the lighters, the narrow, flat-bottomed cargo boats used on the Lagan Navigation, across the lake. The third house is the former lock-keeper's house adjacent to the 27th Lock, sometimes known as the Lake Lock. Across from the lock-house is a small island not too far from the lakeshore, called Tanpudding Island.

Kinnego Marina is the penultimate shore-access point before you return to nearby Oxford Island. The marina is popular with boating enthusiasts and can accommodate up to a hundred boats.

County Fermanagh

Lower Lough Erne
Upper Lough Erne

In her book Prospect of Erne, Mary Rogers remarked that it used to be said that for one half of the year Lough Erne is in County Fermanagh, and that for the other half County Fermanagh is in Lough Erne. While flood-control programmes have diminished the winter enlargement of this extensive lake, a cursory glance at a map of the area reveals a complex, watery landscape dominated by the lake, the River Erne and the many channels of its tributaries. All told, one third of the county is under water. The lake almost bisects the county in a near-perfect diagonal running from its north-eastern boundary with Donegal to its south-eastern boundary with Cavan.

The county's navigable waterways were popular highways for travel and trade from the earliest times, when people were more mobile on water than on land. The shores of Lough Erne and its numerous islands were host to many monastic settlements, with as many as twelve early church sites recorded near the lake. The waterways were also important strategic arteries during tribal feuds and were frequently used for waterborne cattle raids. In more modern times, the broad expanse of Lough Erne proved a viable westerly base for the flying boats that were used to patrol the North Atlantic during the Second World War.

The county name is an anglicised version of the Irish 'Fir' or 'Fear Manach', translated as 'the men' (or 'man') 'of Manach'. The Manach were a small tribe who originally hailed from the province of Leinster and are thought to have settled along the shores of Lough Erne in the second century. The county was ruled by the Maguire clan from around the middle of the thirteenth century until the start of the seventeenth century, when their lands were confiscated as punishment for their participation in the Flight of the Earls. These lands were subsequently handed out to English and Scottish planters (or 'undertakers', as they were then described). The more prominent planters built substantial fortified houses or castles around the shores of Lough Erne. Several of these were subsequently attacked and burned by Rory Maguire as part of a general uprising which took place in 1641 in an effort to recapture forfeited lands. Today, these survive as ruins dotted along the lakeshore; they will be described in more detail below.

The origins of the lake's name are shrouded in legend. Mary Rogers has suggested four possible sources. From the Annals of the Four Masters comes the story of a sept of the Firbolgs called the Ernai, over whom the lake flowed after they did battle on the plain where Lough Erne now lies. In another legend, the name is attributed to a waiting maid of the powerful Queen Maeve of Connacht called Ierne, who drowned in the lake after fleeing from the savage Olcai. Yet another legend from Celtic mythology suggests that the lake is named after a daughter of the Celtic sea god, Manannán Mac Lir. The story told in the final legend, which Mary Rogers attributes to Reverend Henry Newland, will become a familiar one to readers of this book. The legend tells the tale of a fairy spring located down a dark hollow; the spring's waters were covered by a stone and never changed depth or temperature as long as they remained untouched by the beams of the rising sun. On one auspicious morning, two young lovers visited the well to draw water. The boy rolled back the stone from the mouth of the well and dipped their pitcher into the water. Unfortunately, they forgot to replace the cover, and when they were climbing out of the hollow, they looked back – to see the waters boiling up from the spring and flooding the valley below them.

The town of Enniskillen divides the upper and lower sections of the lake; the sections are linked by the River Erne. The upper and lower loughs are vastly different in profile. The lower lough is well defined, and open and very deep to the north, with shallower waters and rocky areas to the south. It is also known as the Broad Lough, and has a complement of ninety-seven islands, some of which are fairly large. The biggest is Boa Island, which stretches across the lake's northern shore and is linked to the mainland by road. Its western shore lies close to the dramatic escarpment of the Magho cliffs. The upper lough could not be more different. It is difficult to say where the river ends and the lake begins. It is much less defined and narrower in profile, and despite the fact that it has fewer islands, at fifty-seven, it has the appearance and feel of having many more. Low-lying drumlins are a dominant feature, with the only prominent landmark being Knockninny Hill, on its western shore. Its largest island, the aptly named Inishmore, is also linked by road to the mainland. Apart from this island, it is possible to cross the lake by road via the smaller Trasna Island.

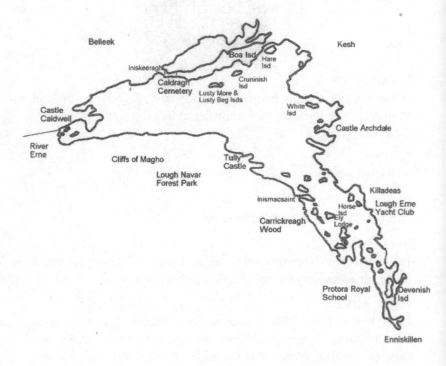

Lower Lough Erne

(98 kilometres)

LOCATION
In the heart of County Fermanagh to the west of Northern Ireland, stretching from the town of Enniskillen at its most southerly point to the Roscor Viaduct.

LENGTH
29 km/18 miles

WIDTH
18.8 km/11.5 miles at its widest point

AREA
15,120 hectares (Note that this is the maximum area; there is considerable variation between winter and summer levels.)

PUBLIC ACCESS
East Shore – Trory, Killadeas, Rossclare, Rossigh, Castle Archdale Country Park, Muckros and Drumrush. Northern Shore – Kesh pier, Lackboy on Boa Island, Rosscrennagh, Rossharbour bay, Garvary, Castle Caldwell. West Shore – Ardees Lower looking on Muckinish Island, Magho jetty, Shean jetty, Drumcrow East beside Tully bay, Camagh near Inismacsaint. There are also a number of places along the lakeshore yielding informal access.

MAP
Ordnance Survey of Northern Ireland Discoverer Series Map 18 covering Enniskillen (ISBN 978-1-873819-58-6) and Fermanagh Lakeland Outdoor Pursuits Map and Navigation Guide for Lower Lough Erne (ISBN 978-1-873819-34-0)

Enniskillen

The town of Enniskillen will be the starting point for the cycling circuits of both the lower and upper sections of Lough Erne. The town's
name is the anglicised version of the Irish 'Inis Ceithleann', which
translates as 'the island of Ceithle'. There is some confusion as to the
identity and gender of Ceithle. According to the Annals of
Clonmacnoise, Ceithle was a king of the Tuatha Dé Danann, the
'divine people' of Irish tradition. Another strand of legend suggests
that Ceithle was female and was one of a number of warlike women
who featured in Celtic mythology. It was said that she was the wife of
Balor of the Mighty Blows, a one-eyed chieftain belonging to the
Fomhoire, a race of demonic beings who occupied Ireland prior to
being defeated during the Second Battle of Magh Tuireadh (or
Moytura, to use its modern name), which is located near Lough Arrow
in County Sligo (see page 167). In a complete reversal of the earlier
legend, Ceithle, rather than being a member of the Tuatha Dé
Danann, is said to have slain one of its principal members, Dagda, at
that battle. Such are the wondrous mysteries of Celtic mythology!

The town was established on an island of 27.5 hectares, though it
has since developed spurs that reach out to the mainland on both sides
of the River Erne. Its foundation is attributed to Captain William
Cole, who was installed as Constable following the Flight of the Earls
in 1607. It was previously a stronghold of one of the lesser branches
of the Maguire clan, whose main base was in Lisnaskea, further north.
A Maguire called Hugh the Hospitable, who died in 1428, built a castle on the island to defend the waterway between the upper and lower
loughs, and the distinctive watergate located on the edge of the river
was added around 1580 by Cú Chonnact Óg Maguire, who died in
1589. Because of its pivotal strategic position, the castle was the target
of attack in the late sixteenth century and early seventeenth century,
and it changed hands a number of times before finally falling to the
English in 1607. The castle was extensively remodelled by Captain
Cole and in the late eighteenth century became an artillery barracks. It
is now a museum, and worth a visit.

Perhaps because of its island location, which made it difficult to
attack, Enniskillen Castle survived unscathed both the general rising in
1641 and the Williamite Wars of 1688–90. The townspeople did, however, play a major role in the latter conflict. A regiment from the town

was at the forefront of the force led by King William III of England, better known as William of Orange, in their final charge at the Battle of the Boyne in 1690. The town has a unique place in military history in that it is the only town in the United Kingdom to have raised two regiments: the 27th, or 'Enniskilleners', and the 6th, or 'Enniskillen Dragoons'.

The Cole family

The Cole family has been closely associated with the town since its foundation by Captain Cole, who was knighted in 1629. The family was ennobled in 1760, becoming Barons of Mount Florence and, in 1789, of Enniskillen. It is hard to miss one of the town's principal landmarks, Coles Monument, which is located in a prominent wooded park at Forthill. The monument was erected in 1843 to the memory of General Sir Galbraith Lowry Cole of the 27th Regiment. The family eventually settled in Florence Court, built in the middle of the eighteenth century by the then Earl of Enniskillen, John Cole. This extensive house, located 12.5 km south-west of Enniskillen on the Sligo Road, was almost destroyed by fire shortly after it had been handed over to the National Trust by the Earl of Enniskillen in the 1950s.

It will come as something of a disappointment that the route along the lake's south-eastern shore yields only sporadic views of the lake unless you venture down some of the dead-end lanes and roads that lead to the lakeshore. It is not until you reach Castle Archdale that you will be in regular sight of the lake. Some of the dead-end roads are worth going down, and these will be signalled as we progress.

Devenish Island

To exit the town for an anticlockwise circuit of the lake, you will have to take the Cornagrade Road to the roundabout, where a northbound exit leads on to the A32 in the direction of Kesh. Just before the junction with the B82, a turn to the left will lead you down to the jetty at the southern tip of Trory bay, from where there are good views of one of the most important of the lower lough's islands, Devenish Island. Boat tours aboard the MV Kestrel are available from Enniskillen. The tour includes thirty minutes on the island; if a longer stay is desired, there is also a ferry facility at Trory jetty.

The name Devenish is from the Irish 'Daimh-inis', meaning 'Ox island'. The island has been uninhabited since 1922 and is now in the hands of the State. Its ruins are testimony to a rich history, mostly ecclesiastical. There were previously more ruins, but in less enlightened times the stone was plundered and shipped for use elsewhere in other building projects. The following quote from William F. Wakeman in 1874 pulls no punches: 'Perhaps there is not besides in all Ireland a scene of such utter wantonness of dilapidation and desecration as shocks a thoughtful visitor to Devenish.'

The earliest structures on the island relate to St Molaise, whose name means 'little flame'. St Molaise was a native of Carberry, County Sligo, and was reputed to have been of royal descent. He studied under St Finian at Clonard and founded a monastery on the island in the sixth century. It is not known exactly when he arrived on Devenish Island, but he is believed to have died between 563 and 571. The ecclesiastical structures consist of the remains of a small church now known as St Molaise's House. It is believed to date back to the twelfth century and is possibly a reworking of an earlier structure. The island's most valuable treasure is a book shrine, the Soiscéal Molaise, which is thought to have been made between 1001 and 1027. It consists of a small bronze container, plated in silver and with gilt patterns, and is on display in the National Museum in Dublin. A handbell which may have been used in connection with the island's round tower is another of the island's treasures to be found in the National Museum. The island has a small museum of its own, which features displays of carved stones and other artefacts related to the island.

Perhaps the most interesting structure on the island is the well-preserved round tower, complete with its conical cap. Within the tower, there are five floors above an unlit basement. Though not as tall as other Irish round towers, reaching a height of only 25 m, it is unique in that it is the only tower with significant carved features that are unrelated to the structure's doorway. Around the base of the cap there is a decorated cornice with a carving of a human head over each of the four top windows. Three of the carvings are of men with plaited beards, while the head over the north-facing window is that of a woman or a young man. Round towers were first built in Ireland in the late nine or tenth centuries, as defences against Viking raids, but some were constructed long after the Viking threat had passed. The elaborate style of the carvings on the Devenish tower suggests that it is one of the later towers, built in the twelfth century. Curiously, the remains of what is thought to have been an earlier tower were discovered in 1973 while post-holes were being dug north of the surviving structure. There is an ongoing debate as to whether these predate the twelfth-century tower – perhaps originating as far back as St Molaise's time – or are the aborted works of an alternate site for the surviving tower.

West of the tower, and occupying the highest ground on the island, are the remains of St Mary's Abbey, which was begun by the Augustinians around 1130 and was extended between 1200 and 1360 to accommodate a larger number of residents. It was reconstructed by Matthew O'Duggan in 1449, as indicated in an inscription on the south wall, under the tower. The distinctive quadrilateral tower attached to the church was added in the sixteenth century. The abbey was abandoned in 1603.

Lough Erne Yacht Club

Returning to the circuit, the B82 takes you past St Angelo Airfield, which was opened in September 1941 in association with the military facilities established at nearby Castle Archdale. Further down the road, a sign points the way to Lough Erne Yacht Club. A review of the map would suggest that there is a track that loops past the yacht club and follows the lakeshore around a peninsula, eventually leading back to a shore-access facility to the rear of the Manor House Hotel and then onwards to Killadeas. While you may at this stage be anxious to get closer to the lakeshore, this track should not be taken. The yacht club gates bar entry to non-members, and while there is a pedestrian entrance, you will be entering onto club property, and this should be respected.

Killadeas Stones

Pushing ahead, you pass the Manor House Hotel, which was previously called Rockfield and was the home of the Irvines, one of the prosperous landowning families who gave their name to the nearby town of Irvinestown, located to the east. It was built in 1746, with an elaborate façade being added in the mid nineteenth century. A short step further on is the townland of Killadeas, where the yard of the church to the left of the road contains some curious stones, which are believed to have origins in both pagan and Christian culture. One of the stones, known as the Bishop's Stone, depicts an ecclesiastic with bell and crozier. Another has a curious combination of scooped-out hollows, whose purpose has given rise to much debate.

Castle Archdale

Moving north from Killadeas, the road squeezes between rising ground on both sides, where ringforts were sited in bygone days. The one to the right is identified as Cassidy Ancient Rath; a burial mound occupies the summit. A little further on from the turn for Rossigh bay, you have the opportunity to leave the main road and enter the southern reaches of Castle Archdale Country Park. A visit to this park would justify an excursion in itself, enabling you to delve into the history of the place and avail of the variety of outdoor facilities now on offer. For cyclists, the tracks through the forest form part of the Kingfisher Trail (Route 91 of the National Cycle Network).

The story of this park goes back to John Archdale, an English planter who originally hailed from Suffolk. He was granted 405 hectares – a modest plot in comparison to other planters – and in 1615 built a castle, the impressive ruins of which can be seen to the north of the forest, perched on high ground. It was more like a fortified house, with two square corner towers at one end of a bawn (a fortified enclosure) and a three-storey, towerlike house at the other. The castle was captured by Rory Maguire in 1641 as he tried to recover lands granted to the planters that would have belonged to the Maguire clan prior to 1607. The castle was eventually abandoned in 1689 after it had been burned out during the Williamite Wars.

The Archdale family line survived thanks to the heroics of one of the castle's nursemaids, who saved the life of the heir, William: all the other Archdale children were killed during the attack. The name could have been lost in later years when the only heir was a lady called Angel Archdale. However, when she married in 1728, her husband, Nicholas Montgomery, took the unusual step of taking her name. In 1773, the family built a grand manor house, which unfortunately no longer stands. The courtyard buildings of that house are all that remain: they house a visitor centre, which includes an exhibition on the activities that have taken place in this park over the years. There are also tea-rooms at the centre, which are open during the summer months.

Lough Erne's flying boats

One of the more interesting aspects of Castle Archdale's history is the fact that it was used during the Second World War as a base for flying boats. When you get down to the lakeshore, it is clear why this was the case. Castle Archdale is situated on the eastern extremity of the widest point of Lower Lough Erne, affording the broad-winged, pot-bellied Sunderland and Catalina flying boats plenty of water for taking off and landing. What was their function, and why Lough Erne? They had two jobs. Firstly, they were detailed to search for German U-boats, which were playing havoc with the convoys of Allied ships that were crossing the Atlantic. Their second job was to provide an airborne escort for those convoys. The choice of Lough Erne was simple: it was the most westerly point in the United Kingdom from which aircraft of this type could be put into the air. Having said this, there was a little diplomatic dancing required to facilitate the Allied use of the airspace of the Irish Republic, a neutral country.

The base established at Castle Archdale saw both success and tragedy. Notably, it was involved in the sinking of one of Germany's most celebrated battleships, the Bismarck. The ship, which had been damaged in a naval battle two days earlier, had been spotted by a Catalina from Lough Erne on the morning of 26 May 1941 as it limped its way towards France for repair. The ship tried to prevent the Catalina from shadowing it by showering it with anti-aircraft gunfire, but despite being damaged, the aeroplane continued its monitoring duties. From the information relayed back, the Bismarck was engaged and sunk the following day.

Although there was a daily requirement for patrols, takeoff and landing were sometimes hampered by the weather. In addition, the surrounding terrain was difficult to negotiate for low-flying aircraft that had limited manoeuvreability. As a result, several crews perished on the slopes of the Ireland's mountainous north-western coastline. In January 1943, a Sunderland crashed at a location called 'Troublesome Rock' when the pilot misjudged his take-off. Fortunately, the crew escaped, but the broken aluminium frame of the aircraft lies at the bottom of the lake, covered in zebra mussels.

It is said that, after the war, the United States offered to sell a number of the flying boats to the British, but in the cash-strapped post-war years, the British could not afford to purchase them. The tale is told that, rather than give them away for nothing, the US air force scuttled some of the planes in the Broad Lough, as the widest stretch of the lake is known, in 1947. Attempts to verify this story have been thwarted by the lake bottom's peat-stained darkness, which has been described as a 'blackness that closes in'.

Flights from Castle Archdale continued until January 1957, and the base was formally closed a year later. Parts of the airbase still exist: the concrete foundations for the barracks and service areas now serve as the hard stands for a thriving caravan park, and the Shetland Dock, which was specially designed to accommodate the flying boats, can be seen on the lakeshore near the park's campsite. The land on which the forest now lies was acquired in 1954 from the Archdale family and is managed by Northern Ireland's Forest Service.

White Island

At the northern exit of Castle Archdale Forest, a turn left, located near the old castle ruins, will place you on the C409. Be prepared for a sustained climb, but at the end of it you will be rewarded with magnificent views across the islands that lie in Castle Archdale bay, in particular White Island. This is another of Lower Lough Erne's more famous islands, mainly due to a collection of beautifully carved stone figures, some lifesize, which have been sporadically unearthed on the island and are reputed to be more than a thousand years old. The last of them was found in 1958. They had all been reused as building blocks for the island's church, which was built c. 1200. There are in fact eight carvings on display within the church ruins. Five of the carvings have been interpreted as bearing Christian images. A sixth appears to be more secular and shows a cross-legged figure, possibly female, with a grimacing face. The seventh piece is a plain block with rough markings, which has an unfinished appearance. The eighth is a more simple carving of a frowning face, with a ribbed head-covering; it is thought that this may have been carved at a later date than the others. As is the case with carved stones found elsewhere in this region, there is ongoing debate as to the purpose or use of these stones. Some believe that the carvings may have been intended as supporting corbels, while others are of the view that they may have served as caryatids, supporting an altar.

The island's ruined twelfth-century church has a delightful Hiberno-Romanesque doorway. The island can be visited by ferry from the marina at Castle Archdale at weekends and during July and August.

Boa Island

An undulating course with glimpses of the lake will bring you inland to the town of Kesh, from where you turn left on to the A37. A brief cycle northwards will take you to the junction with the A37, which will set you on the path for Lower Lough Erne's largest island, Boa Island. The island is just under 8 km long and 1.6 km wide and is connected to the mainland by two concrete causeways that were built in 1925/26. Its curious name comes from the Irish 'Badhbha', a warrior goddess in Celtic mythology who specialised in destruction and creation. It has been translated as meaning either 'carrion crow' or 'battle fury'; perhaps the latter translation is more appropriate, given the interests of the goddess.

The island is most famous for its Janus or Janiform figures, which are to be found in Caldragh Cemetery, near its western edge. Janiform figures are two-faced, with images facing opposite ways on the same stone. Before I visited Caldragh Cemetery, I had read quite a bit about these stones and was looking forward to inspecting them at close quarters. I found the inspection somewhat disappointing. In my view, the figures did not live up to their publicity, perhaps because of the state in which I found them. For such important artefacts, believed to date back to pre-Christian times, their site and condition leave a lot to be desired. The cemetery is located about 300 m from the road, close to the lakeshore, and can be accessed using a paved track that is shared by the adjacent farmyard. I am not sure whether it is a policy decision to leave the graveyard in as natural a state as possible, but it has a generally untended appearance, being more like a rockery than a graveyard, and the two important stones lie almost in the centre, exposed to the elements. Etched on the stones are large triangular faces with prominent oval eyes and mouths above disproportionately small bodies. The smaller stone was actually found on nearby Lustymore Island.

There have been many theories advanced as to the origins and purpose of the stones. Some have suggested that they are fertility idols, while others point to the importance of the head in pagan times as a symbol of the sun.

One would imagine that, as you cross a fairly narrow island like Boa Island, you will enjoy good views of both the southern and northern parts of the lake. That is not the case, however: rising ground to the north of the island obstructs the view in that direction until you are close to the western edge of the island, while high trees and hedgerows along the roadside inhibit the view to the south. Lying close to the southern shore is Lustybeg Island. This is the largest inhabited island on the lake. It is not connected to the mainland by a bridge or causeway but can be accessed by a small ferry. The island is indicated as private but there is a bar, restaurant and bed-and-breakfast facility open to the public.

Castle Caldwell

Iniskeeragh, whose name means 'island of sheep', is wedged between Boa Island and the mainland. You cross it before linking up with the A47, to continue the journey along the lake's north-western shore. The road leads past Castle Caldwell, whose grounds are now an attractive forest and nature reserve, and worth a visit. This former plantation site was previously called Castle Hassett, after Sir Edward Blennerhassett. Sir Edward was the original planter who was granted the lands; he built the first castle on the site in 1612. The Caldwells became the owners of the property fifty years later. The castle is now in ruins and can be reached down a tree-lined road from its entrance, which is flanked by the ruins of an extensive gate lodge. The building is in a terrible state, with trees growing inside it, and ivy and other climbers covering its walls. Signs warn about the dangers of venturing near the building. The area is eerily quiet, but you cannot dispute the beauty of its location, at the base of a forested peninsula. It is a pity that the castle ruins are in such a poor state of preservation. This is probably due to the fact that the ruins and the demesne are owned by the Royal Society for the Protection of Birds, whose agenda is obviously different from that of either the Environment and Heritage Service or the Forest Service. The latter agencies are to be commended for their sensitive preservation and development of some of the other plantation castles and demesnes around the Lower Lough Erne lakeshore. Having said that, it would be wrong to be too critical of Castle Caldwell, because the owners have successfully preserved in a natural state an area that is of striking beauty and appeal. It should be noted that the track that runs to the right of the access road leads to a private road; this prevents a link-up to the Rossmore peninsula and Hare Park.

Belleek

The A47 will take you all the way to Belleek, on the River Erne. Belleek is not part of the circuit but might be considered worthy of a visit by way of diversion. The name of the town is well known worldwide because of its association with fine china pottery. Its name means 'ford-mouth of the flag stone', after the smooth-surfaced stone to be found in the bed of the river nearby. The pottery dates back to 1857 and owes its foundation to a member of the Caldwell family, John Caldwell Bloomfield, who had discovered kaolin (china clay) in the grounds of Castle Caldwell. Even though this supply of kaolin was quickly exhausted, the pottery prospered: the raw materials that produce Belleek's celebrated lustrous finish are now imported.

The River Erne can be crossed by the Rosscorr viaduct, 5 km from Belleek, to start the journey along the western shore of the lake. Here the river is fairly wide and the viaduct has two elements, as if it was built at different times – which it probably was. The first is an iron-and-wood structure, with a raised platform of wooden slats providing the road surface. The second is a conventional bridge with a metalled surface.

The western shoreline of the lake is a more scenic proposition than its eastern or northern counterparts. Even though you will be travelling on a main road with fast-moving traffic, you have the lough in your sights for most of the time, at least north of Carrickreagh. In the northern section, you have the dramatic backdrop of the 305 m-high Cliffs of Magho close by to the west, followed by Lough Navar Forest Park. Added to this are yet more plantation castles to be visited. There are also several places where you can take a well-earned rest, such as at either Magho jetty or Shean jetty, where there are excellent views north to Boa Island and east to Castle Archdale. This would have been a good place to watch the Sunderland and Catalina flying boats making their final approaches to land.

Tully Castle

Tully Castle is sited about 1 km off the main road and is well positioned on a small promontory, from which there are good views of vast stretches of the lake. It was built in 1613 by Sir John Hume, who was one of the principal beneficiaries of the plantation grants; his descendants are also associated with Castle Hume further south. Unfortunately, the Hume family did not enjoy their home for long, as it was another of the castles that was attacked by Rory Maguire in 1641, two years after the death of Sir John Hume. It was left as a burnt-out shell until it received restorative attention in the twentieth century. The ruins have been revived by the Environment and Heritage Service; what distinguishes this castle from the other plantation specimens around this area is that its garden has also been rejuvenated. Visitors can now witness how an Elizabethan garden would have been laid out and enjoy its smells and colours. There is a story associated with the castle that says that Lady Hume hid her jewels close to the castle when it was being attacked by the Maguires. Needless to say, no trace of these valuables has ever been found.

Inismacsaint

Pushing ahead, further south there are fine views of Rabbit Island, which lies off Tully bay, and Inismacsaint, further south. A closer view of the latter island can be obtained from the car park on the lakeshore closer to the island; the car park can be accessed from the main road. This 22.7 hectare island is another important religious site, associated with St Ninnidh. Despite its religious-sounding name, it is an anglicisation of the Irish for 'island of the plain of sorrel'. Unfortunately, there is no regular ferry service to the island. St Ninnidh lived in the sixth century and is also associated with Knockninny Hill, on the western shore of Upper Lough Erne, where he is said to have fasted. The church on the island does not date back to St Ninnidh's time but is an amalgam of different periods, with the earliest construction dating back to the tenth or eleventh century. The most interesting artefact on the island is the cross that stands to the south-west of the church. While it is thought in some circles that it could date back to the eleventh or twelfth centuries, it could very well be a later addition. It appears fairly plain, but it is unusual in that it does not have a radial circle encompassing its arms, as you would find on high crosses such as those at Clonmacnoise, Monasterboice and Kells, usually referred to as Celtic high crosses. A delightful local legend reports that the cross rotates three times every Easter Sunday morning.

As you move further south, you pass through the wooded environs of Carrickreagh. For those who might wish to escape the traffic, you will find a path to the left among the trees. The path, which runs parallel to the road, is located close to the disused Carrickreagh quarry. You may have to lift your bicycle over the fence to avail of this route, and it is important that you emerge at the official entrance to Ely Lodge Forest, as otherwise you could end up wandering along the beautiful forest trails into private land. Be on your guard, as it is easy to miss the boundaries between the public and private land.

Castle Hume

The Humes built several homes along the western shore of Lower Lough Erne. The first of these was Tully Castle, described above. Further south, in the areas of Drumcose and Ross, the family built several homes that do not survive today. One of these was Castle Hume, which was built around 1727 or 1729, to the design of Richard Cassels, the German architect who later designed Leinster House in Dublin and Powerscourt House in County Wicklow. It is said that Castle Hume was the first house Cassels designed in Ireland. Sadly, there is no trace of this building now, but it is said to have been very much like Hazelwood House on the shores of Lough Gill (see page 178). In the early years of the nineteenth century, the house was demolished; its stones are reported to have been used in the construction of Ely Lodge, built on nearby Gully Island. That house did not last long either, and it was blown up in mysterious circumstances in 1870. The current Ely Lodge, which is private, is an extension and modification of the stable block that used to be attached to the original building. Castle Hume now lends its name to the fine golf club that you pass as you progress south.

Portora Royal School and Portora Castle

As you approach Enniskillen once again, you pass the imposing front gates of Portora Royal School. The present school building, an impressive Georgian structure, was erected in 1777, although the school itself was established in the previous century. Two of its most famous past pupils are the literary giants Oscar Wilde and Samuel Beckett. A plaque in memory of Beckett was unveiled on the front of the school on 31 May 2006, a few feet away from a similar plaque commemorating Wilde, who was honoured earlier. It is perhaps strange that two of the stars of Irish literature, who both hailed from Dublin, should have spent a good deal of their schooldays in Enniskillen. While Wilde features on the school's honours board for academic achievement, there is no mention of Beckett. It appears that he concentrated on athletic endeavours, excelling at rugby, boxing and cricket. In the vicinity of the school is Portora Castle, which dates back to 1612 but fell into ruins in the nineteenth century. The Cole family lived in this castle while they were waiting for their new home at Florence Court to be completed.

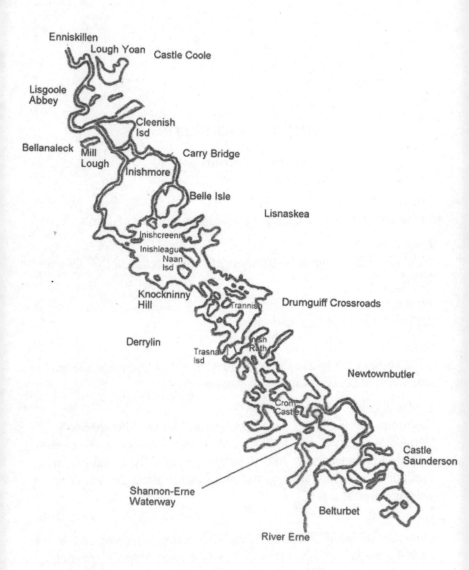

Enniskillen
Lough Yoan
Castle Coole
Lisgoole Abbey
Cleenish Isd
Bellanaleck
Mill Lough
Carry Bridge
Inishmore
Belle Isle
Lisnaskea
Inishcreenm
Inishleague
Naan Isd
Knockninny Hill
Trannish
Drumguiff Crossroads
Derrylin
Trasna Isd
Inish Rath
Newtownbutler
Crom Castle
Castle Saunderson
Shannon-Erne Waterway
Belturbet
River Erne

Upper Lough Erne

(*c.* 90 kilometres)

LOCATION
In the southern half of County Fermanagh, south-east of Enniskillen

LENGTH
14.5 km/9 miles from the south shore of Inishmore Island to Galloon Island

WIDTH
4.5 km/2.8 miles at its widest point

AREA
5,772 hectares (Note that this is the maximum area; there is considerable variation between winter and summer levels.)

PUBLIC ACCESS
East shore – Dolan's Ring, Carry Bridge and Marina, Kilmore South, Corradillor, Derryad and Belturbet. West shore – Geaglum, Tiraroe, Cornaleck, Knockninny and Bellanaleck. There is a marina at Enniskillen and there are also numerous places along the lakeshore yielding informal access.

MAP
Ordnance Survey of Northern Ireland Discoverer Series Map 18 covering Enniskillen (ISBN 978-1-873819-58-6) and Map 27 covering Upper Lough Erne (ISBN 978-1-873819-86-9)

Designing a cycling circuit for Upper Lough Erne is difficult because of the lake's scattered profile and the debatable issue as to where it actually starts and finishes. For the purposes of this book, the circuit will embrace the area between Enniskillen and Belturbet, even though both of these towns are not strictly on the lakeshore. They are, however, connected by the River Erne. While the circuit shown is one complete trail, the existence of two crossings allows for subdivision should the reader be so inclined. The first crossing is facilitated via Inishmore Island, while the second uses Trasna Island, further south. The lake's scattered profile means that for some lengthy stretches, maintaining close contact with the lake will not be possible and views of the water will be limited. However, there is ample opportunity to extend the route so that you travel close to the lakeshore as you progress around the circuit.

Starting out from the car park adjacent to Enniskillen Castle for a clockwise circuit, it is recommended that you follow the signposted trail for the National Cycle Network Route 91 for the early part of the journey through the outer suburbs of Enniskillen. This includes a fairly strenuous climb out of the town, leading to the Ardhowen Theatre and Arts Centre and on to the busy A4.

Castle Coole

The entrance to the landscaped grounds of Castle Coole soon appears to the left of the A4. This is a magnificently preserved Palladian mansion that was built between 1790 and 1798 and is now owned by the National Trust. It was the family home of the Earls of Belmore and was built by the first Earl, Armar Lowry, who died in 1802. He was a descendant, through the maternal line, of the Corry family. To call this building a castle is a misnomer, but it inherited this appellation from previous structures that were erected on the grounds by the Corrys and the first plantation owner, Captain Roger Atkinson, in the seventeenth century. The first Castle Coole was burned down by its owner, James Corry, in July 1689 so that it would not fall into the hands of the supporters of King James II during the Williamite Wars of that period. He was later compensated by King William III when he succeeded to the throne, allowing him to rebuild the castle in more grandiose style. If you are visiting this area during the winter, you will not have the opportunity to take a tour of the house: like other National Trust properties, it is open on a seasonal basis. However, you can still wander through the wooded estate which overlooks Lough Coole.

A short cycle south of Castle Coole is the first of the possible extensions to this route. A turn to the right leads past Lough Acrussel to Dolan's Ring, where there is a pleasantly sited picnic area on the bank of the river, looking over towards Bellanaleck on the western shore. The picnic area is just over 3 km from the turn-off from the A4; the extension would add 6 km to your circuit.

Inishmore Island

Returning to the A4, you finally get the opportunity to leave that busy road by taking the B514 in the direction of Carry Bridge and Marina. This is located at the eastern end of the upper lough's largest island, Inishmore, which is also joined to the mainland at its western end by the Inishmore viaduct. The island effectively marks the northern boundary of the upper lough. It is large, stretching more than 4 km along the northern shore of the upper lough, and within its confines it possesses several of its own lakes, such as Lough Barry. Despite the fact that the island is low-lying, a crossing by road yields few views of its northern shoreline, or of Upper Lough Erne to the south. Surprisingly, the island has no monastic history, despite its size and pivotal location.

Belle Isle and the Annals of Ulster

Immediately south of Inishmore is the island of Belle Isle, which has a far more interesting history. The island is clearly signposted off the road south from Carry Bridge which runs to the western side of Derryhowlaght Lough. The island is in fact two islands separated by a narrow waterway and connected by a bridge. Ballymacmanus Island is nearest the mainland, and West Island is further out. It was here that the Annals of Ulster were compiled in the fifteenth and sixteenth centuries. These important texts are one of a series of annals that recorded the history of Ireland back to prehistoric times; the annals have been pored over, analysed and drawn upon by modern historians for many years. (Reference will be made later in this book to the Annals of the Four Masters and the Annals of Loch Cé, both of which have lakeshore connections.) The Annals of Ulster owe their existence to a learned Irish chieftain, Cathal Óg Mac Manus (1439–98), and take the form of two texts. The last entry in the first text is dated 1504; this text now lies in the library of Trinity College Dublin. A later text, in which the last entry is dated 1588, is housed in the Bodleian Library of Oxford University. The texts were written in a mix of ancient and modern Irish, with some parts in Latin. As far as the texts are concerned, Irish history began with the mission of Palladius to Ireland in 431, the year before St Patrick arrived on our shores. It is believed that the Four Masters used the Annals of Ulster as a reference when compiling their more extensive work in the late seventeenth century.

Belle Isle Castle is a sprawling estate owned by the Duke of Abercorn, who bought it in 1991. The castle is hired out for weddings and for long weekends.

Lisnaskea

Returning to the mainland, you can continue south on quiet but undulating country lanes that run parallel to the B514 before linking up with that road at Ballindarragh Bridge, which spans the Colebrook river. A brief cycle takes you to Lisnaskea. Though not directly on the lakeshore, this town was important in the region, as it was the base for the main branch of the Maguire clan, as mentioned earlier. The ruins of a fine plantation castle stand in a prominent position to the left of the B127 as you exit the town to the south. Known as Castle Balfour, it was built for Sir James Balfour around 1620. As with the other plantation castles in this area, it succumbed to the usual pattern of attack and repair during the 1641 uprising and the Williamite Wars of 1688–90. In 1780, the Balfours left Fermanagh; the castle passed into the hands of the Crichtons of Crom, before being destroyed by fire in 1803, after which it was abandoned. The derelict ruins were taken over by the State in 1960, and a partial restoration was carried out in the following decade. Today, access to the castle is through the graveyard attached to a church that stands on what would have been part of the original castle grounds. An information panel at the site illustrates how the castle would have looked in its heyday.

After a featureless 4 km ride south out of Lisnaskea on the B127, you have the opportunity to link up once again with Route 91 of the National Cycle Network at Drumguiff crossroads, close to Lough Doo. This junction is important to cyclists. A continuation on the

B127 will lead you to a crossing to the western shore via Trasna Island, trasna being the Irish for 'crossing'. The crossing is part of both the Ulster Way and the Kingfisher Cycle Trail and will be of interest to those who might find the full circuit of the Upper Lough too onerous. Those following the Kingfisher Cycle Trail might wish to note that during the summer months there is also the possibility of crossing by ferry further south at Crom Castle.

Inish Rath and the Hare Krishnas

From Drumguiff crossroads, follow the Route 91 signposts south. Where this trail branches off to the left at the next junction, there is the opportunity of making a small extension to the route by continuing straight, going down to the lakeshore opposite Inish Rath, an inhabited island that is interesting because of its present-day occupants. The island, which was one of the homes of the Butler family until the start of World War II, has been home to a settlement of Hare Krishna followers. It is interesting that they should follow in a long line of island dwellers on Irish lakes focused on the practice and dissemination of religion.

Returning to the circuit, it is somewhat surprising that Route 91 progresses inland rather than enjoying a scenic route that courses between a number of appealing lakes that break up the undulating countryside. Perhaps it is the deterrent of some stiff climbs and descents. The route I recommend requires you to turn right just past Derryanny Bridge and then bear left almost immediately, followed by another sharp turn right. This extremely quiet but hilly road brings you past a series of five fine lakes, all of which are identified along the road, before again linking up with Route 91 close to the road that leads to the entrance to Crom Castle.

Crom Castle

Crom Castle is one of the highlights of this circuit and is welcoming to cyclists. Route 91 actually runs through the 809-hectare estate, which is owned by the National Trust. A word of warning for those who might contemplate a visit during the winter months is that the grounds are only open on Sundays during the winter and the visitor centre is only open during the summer season. In addition, the ferry that operates from Crom to the western shore at Derryvore is only available during the summer months.

There are two castles on the estate, which occupies a strategic point along the water highway between Enniskillen and Belturbet. The original castle lies close to the lakeshore and was built between 1610 and 1611 by Michael Balfour, Laird of Mountwhinney. Unlike most of the other plantation castles, it survived the troubles of the seventeenth century relatively unscathed but was destroyed by an accidental fire in 1764. Its present-day ruins are not all that they may seem. They were embellished in the nineteenth century to make them look more romantic, with the addition of walls and towers on either side of the central structure.

The family that has been most associated with Crom are the Crichtons, who hailed from Brunston in Scotland and, when ennobled, took their title from the lake. The present Lord Erne, John Crichton, who is the largest landowner in County Fermanagh, still lives on the estate, in the more modern neo-Gothic castle, which was built in 1834. This castle is not open to the public.

There is so much to see on the Crom estate, and it is so well presented, that you come away with an excellent understanding of the enterprise that was required to run an estate of this type. The visitor centre, which is housed in the estate's former farmyard, hosts an exhibition that outlines the working of the estate. Around the grounds you will find information panels which describe the purpose and operation of the various buildings to be found on the estate, such as the ice house, the turf house and the boat house. There is so much to see and absorb that your visit to the Crom estate will be a lengthy one.

There are a number of interesting islands lying close to Crom estate. Looking out from the visitor centre is Inishfendra. The translation of this name from the Irish is 'island of the fair rath'; this relates to the remains of a Maguire castle which is located at the eastern end of the island. The closest island to Crom is Inisherk, which has been linked to the mainland by a bridge since the nineteenth century. The current bridge, known as the White Bridge, after its predecessor, was installed in 1961. On this island are a number of estate buildings, and also a walled garden that is not open to the public but which used to supply the estate's needs for fruit and vegetables. On the opposite shore, you will be able to see Holy Trinity Church, which is sited on Derryvore peninsula – already mentioned as the landing point for the

Route 91 ferry.

Leaving Crom estate behind, take the road towards Newtown-butler. A turn right about 1 km east of the town enables you to progress south in close proximity to a number of smaller lakes. You link up with the B533 just north of the old course of the Ulster Canal, which used the River Finn to access Lough Erne east of this point. You cross the River Finn itself a little further south, before crossing out of Ulster and Northern Ireland and into County Cavan and the Republic, although you will not notice any boundary or frontier noti-fication.

Castle Saunderson demesne

Once you cross the border, the road, now called the N54, swings to the left. At the first opportunity, you should turn right on to a second-ary road, where almost immediately you pass one of the entrances to Castle Sauderson demesne. This demesne is now under the manage-ment of Coillte, the Irish Forestry Service, and is a working forest. Castle Saunderson was one of the three great estates in this area, and rivalled Crom and Lanesborough Lodge, the home of the Butlers. When Ireland was partitioned in 1922, the Saundersons found that their estate was located in the Republic, despite the fact that they would have regarded themselves very much as being loyal to the United Kingdom. Cut off from their natural associations, they gave up their mansion, which now lies in ruins on the banks of the River Finn and can be inspected by walking through the demesne. In contrast with the position across the border, there is no signage indicating the presence of these ruins, or even the existence of the demesne itself. This is regrettable, as the ruins are delightfully sited and the demesne is a rich woodland worthy of exploration.

Moving away from Castle Saunderson, it is possible that the road that heads for Belturbet is the one that used to be known as 'starva-tion lane'. This was the name given to a road built by the Saundersons as a famine-relief project during what is often referred to as the Great Hunger of 1846–51. The Saundersons appear to have been a decent lot. In Prospect of Erne, Mary Rogers notes that, in addition to building the famine road, they did not charge their tenants any rent during the famine period.

As you approach Belturbet, there is the opportunity to take a small loop to the north of the approach road as far as Quivy. A word of warning, though. The Ordnance Survey Map (Sheet 27) indicates a road that reaches right up to Quivy via Quivy Bridge and then moves westward past Killylea and Edenterriff Loughs before returning southwards. Unfortunately, I came to a dead end around Bunanumery and ended up in the driveway of a private residence. Neither I nor the dogs of the residence were too happy. You can still venture as far as Quivy and on your return take the first turn left; this will link up with the road that brings you to Creeny Bridge on the eastern outskirts of Belturbet.

Belturbet

Belturbet stands at the head of the navigation of the River Erne; the town's history has been determined by its pivotal position. Settlements in the area have been dated back to the eleventh century. On Turbet island, upriver from Erne Bridge, you can inspect the remains of a motte and bailey that was constructed by the Anglo-Norman John le Grey in 1210. However, what was once a delightful and proud historical site is now far from enticing. There is a visitor centre based in the old railway centre, but there is little enough to detain the casual visitor.

Shannon–Erne Waterway

The journey northwards begins almost as soon as you have crossed Erne Bridge. Unfortunately, the early stages have to be undertaken on the busy N3. There is, however, a good hard shoulder on this road, allowing for an element of comfort and relative safety. The border crossing into Northern Ireland is easy to spot. It comes at the Senator George Mitchell Peace Bridge, which was constructed as recently as 1998 and named in honour of the US senator, in recognition of his contribution to the Good Friday Agreement of that year. It spans the River Woodford, which is the final element of a waterway that links Upper Lough Erne with the River Shannon, aptly known as the Shannon–Erne Waterway. This is the modern name given to what used to be known as the Ballinamore–Ballyconnell Canal – which in turn was once part of a navigation system that linked Belfast with Limerick for a short time in the nineteenth century. The canal was completed in 1858 but was a commercial failure, closing to traffic nine years later. Its existence was under threat for many years, even as a physical remnant of past heritage, but thanks to an initiative that arose out of North–South co-operation, it has been restored and stands proud as one of the most modern waterways in these islands. After four years of work, it was reopened for navigation in May 1994 as a beautiful waterway, all of 63 km long. A significant part of its meandering course forms the boundaries between Counties Cavan and Leitrim, and also Fermanagh and Cavan. As evidence of its modernity, it uses automatic locks operated by key cards. It is also an ideal route for cycling, although the towpaths that once bordered the lakes through which it passes no longer exist. In these areas, the waterway can be tracked easily by road.

If you wish to track the canal on its final passage towards its destination, there is a secondary road that runs parallel to the winding course of the waterway virtually all the way to the Erne, opposite the island of Inishfendra. Cycle on about 1 km to the north of Senator George Mitchell Peace Bridge, and you will see the road branching off the A509 to the right. The road runs between several lakes on both sides, and the waterway is never far away. It ends at a quaint, recently derelict cottage, after which there is a track behind a gate; this track leads all the way down to the shore of Upper Lough Erne, just north of the junction with the waterway. As the track is on private land, permission should be sought.

During the winter months in this area, you are likely to see whooper swans, who come to escape the harsh Icelandic winter. The reed-fringed edge of Lough Erne is a favourite destination for some of the migrants, and indeed many can be seen grazing in nearby fields. In fact, it is estimated that as many as 5 percent of the world's population of whooper swans winter here. The swans are the beneficiaries of a government scheme in Northern Ireland that encourages farmers in the area to allow these birds to graze on their land undisturbed. Farmers are paid to take cattle off the fields early in order to accommodate the arrival of the swans, who would otherwise compete with the cattle for grass. The land must be set aside for swans until 1 April, by which time they have returned to Iceland.

Knockninny Hill

Returning to the A509, you will shortly have the opportunity to continue the circuit for an extended period on the quieter secondary roads
and country lanes that will bring you closer to the lakeshore. Take the
second turn to the right, and after a little more than a kilometre you
will link up once again with the signposted Kingfisher Cycle Trail.
Should you miss this turn, do not worry, as the road to the right at the
Teemore village crossroads will also connect you to this trail. Follow
the Kingfisher signs towards Derrylin, passing by the western shore of
the Trasna Island crossing point. Just before you reach the village, a
turn right will have you heading back towards the lakeshore for a
delightful loop around Knockninny Hill. I am once again surprised
that the designers of the Kingfisher route took the route so close to
an area that presents some of the best lakeshore views along the western shore but instead chose to omit it.

There is a public-amenity area on the lakeshore at Knockninny,
and also a private marina operated by a cruiser-hire company. Once
you have passed the marina, the road rises past the northern side of
Knockninny Hill, which takes it name from St Ninnidh, who is said to
have fasted there. As you climb, take a moment to look to your right
and back over your shoulder, where you will have splendid views of a
number of well-wooded islands, including Inishlirroo ('Half-red
Island') and Naan ('Ring-shaped'). Late in the year, when leaves adopt
their autumnal colours, one might easily think that there is sun perma-
nently shining down on these islands, even on the greyest of days.

As you skirt the northern side of Knockninny Hill, you cannot
help but notice the quarry which has carved out a huge slice of the hill.
The sheer sides of the quarry face, and the makeshift road ascending
to the top, are awesome. A significant part of the hill has been
devoured, and one wonders how long it will be before its middle section completely disappears, leaving two strangely shaped towers.

Unfortunately, the remainder of the journey back to Enniskillen
has to be negotiated on the A509, with few views of the lake and little of interest in the surrounding hinterland. About 3 km south of
Enniskillen, you pass the entrance gates to Lisgoole Abbey, a former
Augustinian and later Franciscan religious site, which is not open to
the public.

County Cavan

Lough Gowna

Lough Sheelin

Lough Oughter

Lough Ramor

Cavan occupies a pivotal position in the Irish landscape, providing a meeting place for three of Ireland's four provinces. It was once ruled by the powerful clan of the O'Reillys of Breffni (sometimes spelt as 'Breiffne'), the ruins of whose castles are to be found throughout the county, most notably in its principal town and in Killashandra. While it is now part of the province of Ulster, it was part of Connacht up to the sixteenth century, and the case could always have been made for its inclusion in the province of Leinster. Located 112 km north-west of Dublin and 135 km south-west of Belfast, the county is known as 'the Lake Country': it is said to have a lake for every day of the year. It is called after its principal town, whose name is in turn derived from the Irish word cabhan, meaning a hollow.

Cavan can be described as either completely landlocked or totally waterlogged, depending upon your perspective. This combination of land and water, together with the proliferation of drumlins, provides the county and its visitors with a varied landscape that offers much to enjoy. Drumlins are smooth, rounded hills formed by deposits left behind by vast glaciers in the last Ice Age, and Cavan abounds with them. The word 'drumlin' is an anglicised form of the Irish word droim, meaning ridge.

In some parts of the county, there is a great feeling of remoteness, primarily due to its shaded, winding roads and scattered population, and it is an ideal location for the touring cyclist – notwithstanding the undulating landscape caused by the drumlins. It is also hugely popular with anglers, with unlimited bank-space along its lakes, rivers and streams.

It is an important county for any person interested in Ireland's waterways, as two of Ireland's great rivers have their origins here. The River Shannon rises in the Cuilcagh Mountains, to the north-west of the county, and the River Erne has its origins near a small village called Cross Keys, in the east of the county, before flowing due west into the Gowna network of lakes. The Irish name for the village of Cross Keys is Carraig an Tobair, which translates as 'Rock of the Well' – providing a clue as to the spring source of the river.

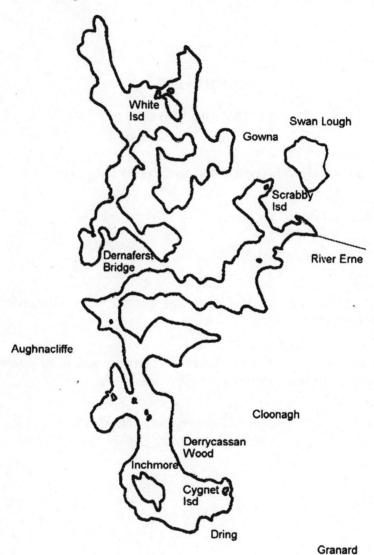

Arvagh

White Isd

Swan Lough

Gowna

Scrabby Isd

Dernaferst Bridge

River Erne

Aughnacliffe

Cloonagh

Derrycassan Wood

Inchmore

Cygnet Isd

Dring

Granard

Lough Gowna

(35 kilometres)

LOCATION
Approximately 6 km north of Granard, County Longford

LENGTH
9 km/5.6 miles

WIDTH
4 km/2.5 miles at its widest point

AREA
c. 1,299 hectares/12.9 km² (Note that the area varies considerably between winter and summer levels.)

PUBLIC ACCESS
Dring, Dernaferst, Cloone, Arnaghan. There are also many places along the lakeshore yielding informal access.

MAP
Ordnance Survey of Ireland Discovery Series Map 34 covering Cavan, Leitrim, Longford, Meath and Westmeath (ISBN 978-1-903974-26-1)

Lough Gowna is shared between the counties of Longford and Cavan, with the former having the slightly longer shoreline. The lake's name is an anglicised version of the Irish 'Loch Gamhna', which translates as 'Lake of the Calf'. According to local legend, an old woman desecrated a holy well by washing clothes in it. The result of this desecration was that a calf jumped from the well, followed by an eruption of water. The calf ran off in haphazard fashion, with the water following it, until it was stopped in its tracks by a local farmer, thereby preventing the water from spreading further. It is not a very deep lake, with the deepest part having been measured at 14.3 m by Frank Columb, author of Lough Gowna Valley.

The circuit around the lake is one of the most rewarding lakeshore cycle routes. The roads have little traffic even at the height of the summer, and low hedges offer clear and sometimes breathtaking views. In addition, there is the potential to avail of several worthwhile extensions, some of which include off-road sections; these could add over 25 km to your circuit in total.

The lake is divided into upper and lower sections, connected at Dernaferst, with the upper portion draining into the lower. There is a road that leads westward from the charming village of Gowna, and for a brief period this route affords a delightful view of the lake on both sides at the one time. A good place to start the circuit is the public lakeshore facility at Dernaferst Bridge, located between the two halves of the lake about 3 km west of Gowna. You will not be able to avoid cycling a portion of this road twice as you complete the circuit, but this is no hardship given the location. Gowna is located approximately 13 km from Granard: you should use the N55 and turn where signposted.

The Northern Shoreline

The choice of covering the northern or southern half of the lake first is a matter of personal preference. This guide opts for the northern course first: go westward from Dernaferst Bridge until you come to a crossroads, where you bear right. This is an inauspicious start, as the road initially runs away from the lake and your view is obscured by hilly ground. A junction at about 3 km from the crossroads will set you back on track, and you will once again be close to one of the two northern extremities of the lake. Watch out closely for the turn to the right, which will take you across the lake's northern shores. The narrow road features several ascents and descents but nothing too taxing and it does not take long to reach the road that links Gowna and the village of Arvagh to the north. The junction you have reached is at the county boundary between Longford and Cavan; in fact, just north of the village of Arvagh is the meeting place of the three provinces of Connacht, Ulster and Leinster.

Gowna

Turning right leads you into the village of Gowna, mentioned above. The village used to bear the unattractive name of 'Scrabby'; it is perhaps not surprising that it was changed! The hill to the north of the town which you passed earlier, and an island on the lake near the village, still bear the names 'Scrabby Hill' and 'Scrabby Island', respectively. Despite its size, the village is well provided for when it comes to public houses. Both Fitzpatrick's Tavern and the Piker's Lodge serve food, while Whyte's is a pleasant stop for a refreshing drink. At the crossroads in the centre of Gowna, turn right, leading back to your starting point. A short distance from the village, there is a large factory on the right hand side; this seems completely out of place in terms of the landscape of the rest of this circuit. As an indication of contrast, there is a quaint whitewashed thatched cottage on the roadside almost immediately adjacent to the factory: progress and tradition standing proudly side by side. On the route back to Dernaferst Bridge there are two access points to the northern part of the lake, firstly at Cloone and later at Arnaghan; both of them are signposted. If you were feeling energetic, a delightful extension to the circuit would be to include the Arnaghan trail, which will add about 5 km to your journey.

Inchmore Island

Bypassing the starting point once again, you make your way to the crossroads on the western shore, and on this occasion turn left. Along this route you will be entertained by some of the best lake views you are likely to encounter in this region. An attractive feature of Lough Gowna are the many small inlets and bays along its shoreline, much of which is lined with a wide variety of trees. Derrycassan Wood is prominent on the far shore. As you progress southward, you will see

Inchmore Island, on which you can observe some ruins. This island, of about forty acres, is Lough Gowna's largest and has been uninhabited since the 1820s. According to local lore, the ruins that you can see are those of a monastery established in 550. It became an Augistinian Priory in the twelfth century and remained so until the dissolution of the monasteries in 1543. The graveyard on the island was used for a long time afterwards, with the last burial having taken place on 16 May 1945. A bell associated with the island, known as the Inchmore Bell, now hangs in the nearby Aughnacliffe parish church.

Derrycassan Wood

At the next crossroads turn left towards the tiny village of Dring, where there is a sizeable car park and shore facility with a jetty. It is easy to get lost in the network of small roads close to Dring. You will find yourself consulting the map frequently and perhaps checking the directions with locals. To ensure that you do not get lost, keep in mind the location of the lake and continue in a generally northerly direction. There are, however, three worthwhile diversions in relatively quick succession that might be included in your travels in this area. The first will take you into Derrycassan Wood, an interesting trail which at present is confined to walkers; the entrance to the wood can be found to the right-hand side of the Dring GAA pitch. The 72.4 hectare wood was once part of an estate owned by the Dopping Hepenstal family, one of the major landowning families in the region of Lough Gowna during the eighteenth and nineteenth centuries. There used to be a large house within the demsene; the house was built in 1760 but was demolished in 1939, and the stone was used to build St Colmcille's Church at nearby Mullinalaghta. You can opt for one of three marked trails through the wood; the longest of these brings you along the lakeshore. If you opt for the longest trail, you will pass the ruins of an old boathouse. This was the location for what has been described locally as 'Lough Gowna's greatest tragedy'. On 3 April 1855, four men – John Dopping and three officers of the Longford Rifles – drowned as they were returning to shore from a party on board Dopping's yacht. A small boat used to carry some of the yacht's passengers to the shore overturned; it is thought that the combination of alcohol and cold water played a large part in the resulting tragedy.

An alternative to the forest trail is to avail of the road that runs parallel to the wood and extends a little further beyond it. Whichever you choose, you will have to backtrack to the original route – although if you are choosing the road, there is a slip-road to the left, about halfway down, which links up with the original circuit a short distance from the access to the third suggested extension in this area. This extension is slightly longer and leads down to another small forest, called Gortaroe Wood, around which is a circular trail that is popular with anglers and walkers. From the shoreline, you will be able to see Dernaferst Bridge a short distance across the lake.

There is a third road off to the left as you pass through a stretched-out village which appears to have two names. 'Mullanalaghta' is inscribed on the welcome sign at the entrance to the village, while the post office is called 'Cloonagh PO'. The map does little to resolve the situation. This road leads down towards the shoreline and deteriorates into rough lane where it leads to some private dwellings. I would not recommend using this extension.

Resuming the circuit once again, it is only a short distance to Gowna, where you will encounter Swan Lough to your right on the fringes of the village. Turning left at the village crossroads will take you down once again to Dernaferst Bridge.

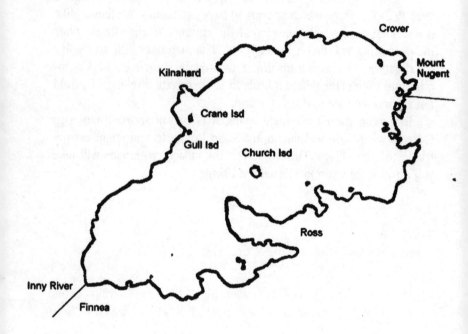

Crover

Mount
Nugent

Kilnahard

Crane Isd

Gull Isd

Church Isd

Ross

Inny River

Finnea

Lough Sheelin

(25 kilometres)

LOCATION
Approximately 8km east of Granard, County Longford

LENGTH
8 km/5 miles

WIDTH
4 km/2.5 miles at its widest point

AREA
1,800 hectares/17.7 km^2

PUBLIC ACCESS
Finea, Kilnihard, Crover, Ross Shore, Tulla

MAP
Ordnance Survey of Ireland Discovery Series Map 34 covering Cavan, Leitrim, Longford, Meath and Westmeath (ISBN 978-1-903974-26-1)

Lough Sheelin is shared by three counties: Cavan, Westmeath and Meath. In the Middle Ages, it provided a border between the Anglo-Normans and the powerful O'Reilly clan. Like many of the lakes in the region, it is a limestone lake; it is also one of a dozen or so remaining brown-trout lakes in Europe that is renowned well beyond its shores. As in the case of Lough Derravaragh, the River Inny courses in and out of the lake. At the south-western extremity of the lake, that river provides a short link between Lough Sheelin and another much smaller lake, called Lough Kinale. Sandwiched between the two lakes is the small, scenic village of Finnea, which is a good starting point for the circuit. The narrow village bridge spanning the River Inny marks the county border between Cavan and Westmeath. The bridge became famous in 1646 when a group of one hundred men led by Myles 'The Slasher' O'Reilly held off over one thousand of Oliver Cromwell's soldiers in a battle that lasted a full day. Local information panels relate the story of how O'Reilly's cheek was pierced by the sword of a gigantic Scotsman and he bit down on the blade of the sword, holding it in a vice-like grip while inflicting a mortal blow on his enemy. A cross on the main street of the village commemorates this deed and O'Reilly's heroism in the battle.

Water-quality problems

Some years back, Lough Sheelin was thrust into the news due to the poor quality of its water. It suffered badly from eutrophication, a complex-sounding word that is best defined as artificial enrichment generated from an invasion of nutrients. The nutrients, the most damaging of which is phosphorous, run off land during rainfall and find their way into our lakes via the streams and rivers that supply them. The overmanuring of water has several detrimental effects. Close to the shore, accelerated rooted-plant growth occurs in the shallower areas, while planktonic algae develop in the open water. This prolific algal development is commonly referred to as 'algal bloom'. The result is that light penetration into the water is reduced considerably, with dire consequences for other plant life, on which invertebrates feed. Fish are also inhibited in their movement and feeding. The lake as a whole suffers from de-oxygenation caused by the death and breakdown of the algae, which also causes changes in the sediment or bed strata on the lake floor. For a long time, Ireland stood alone among European

Union countries in not having implemented an EU Directive which specifically dealt with nitrates and was aimed at alleviating the water-quality problems encountered in our lakes and watercourses. Fortunately, the Minister for the Environment finally signed the directive into law in December 2005; it should help in ensuring that the water-quality problems encountered in the past in lakes like Lough Sheelin will no longer feature.

The cycling circuit around Lough Sheelin is one of the easiest to follow in the book. There is a network of quiet roads close to the lake, providing the cyclist with a pleasant tour with moderate climbs and descents, very little traffic and frequent agreeable views across this attractive lake. Crossing Finea Bridge, turn right almost immediately on to a small country road, which provides a brief respite from the busier R194. Initially, the lake is out of view, and after 1.5 km you link up once again with the R194, bearing right. As the road rises, you catch the first glimpse of the lake. A short distance on from the headquarters of the Lough Sheelin branch of the Shannon Fisheries Board, there is a signposted turn to the right. This leads down to Kilnihard Shore, where there is a small harbour, with many small fishing boats lying at the water's edge. There are also several caravans and mobile homes, which are used to accommodate anglers. There is a good view of two small islands, Crane Island and Gull Island, which lie close to the shore.

Mount Nugent

Returning to the main road, continue to the next crossroads, where you bear right for the journey across the north-eastern extremity of the lake. Not far along this road is the Crover House Hotel and Country Club, whose grounds provide some of the best overall views of the lake. A short ride from the hotel is the public shore facility at Crover, where there is a picnic area, a slipway and a small jetty. The road that follows the lakeshore from this facility is a dead end leading to several private dwellings. The road opposite the picnic area leads to the R154. This will bring you to the village of Mount Nugent, where once again you meet the River Inny as it makes its short way to the lake. The village lies on an old coach road that existed between Granard and Cavan and was originally called Daly's Bridge. It was later renamed after the local landlord. Carraig Crafts Visitor Centre, where the art of weaving is on display, is located just outside the village. Mount Nugent is also home to a successful boatbuilding enterprise, Sheelin Boats, which was founded in the early 1990s by Michael McCabe. Mr McCabe owned and operated the Lough Sheelin Hotel from 1961 to 1990 but was forced to close it down owing to lack of business caused by the extensive pollution in the lake, as mentioned above. The hotel is now run as a nursing home. Sheelin Boats has grown to become the largest family-owned boatbuilder in Ireland.

Turn right at the village crossroads to embark on the journey along the southern shore of the lake, once more on a relatively quiet country road. Tulla Shore is signposted to the right. This shore facility, where there is a slipway and jetty, is located close to the lake's widest point and provides delightful views across the water.

Ross

A steep climb leads across the county boundary between Cavan and Meath into the small village of Ross. A turn right will bring you to Ross Castle, which provides accommodation and from where there are good lake views. The stay in County Meath is brief but there is the chance for a small diversion into a wooded area between the road and the lake. This route can be accessed following the speedy descent out of Ross. This trail, which is confined to walkers at present, leads down to a public shore amenity. Returning to the road, the Meath–Westmeath boundary is quickly passed, and the following stretch of road is a delight for cyclists. For the next 4 km, the road is relatively flat and straight, and extremely quiet, and the landscape alternates between bogland and forest. This section ends with the junction with the R394 leading to Finnea. As you approach the village, there is a signpost for the last shore access at Sailor's Garden.

Belturbet

River Erne

Black Lough

Milltown

Drumlane
Round Tower

Drummany
Lough

Ardan
Lough

Baker's
Bridge

Drumlane
Lough

Urney
Bridge

Inishmuck

Annaloo
River

Garfiny Lough

Carrah
Lough

Deralk
Lough

Corglas
Lough

Derryheen
Bridge

Tullyguide
Lough

Town
Lough

Cloch Oughter
Castle

Inishconnell

Drumard

Killashandra

Mill
Lough

Inch
Isd

Killykeen
Forest Park

Cavan

Trinity
Isd

Tawlaght
Lough

Carr's
Lough

River Erne

Crossdoney

82

Lough Oughter

(35 or 46 kilometres, depending on circuit taken)

LOCATION
Approximately 6 km west of Cavan town

LENGTH
9 km/5.5 miles

WIDTH
n/a

AREA
1,300 hectares/13.0 km^2

PUBLIC ACCESS
Because of its nature, there are many areas for public access. The principal signposted access points are at Killykeen Forest Park, Innisconnell, Drumard and Trinity Island.

MAP
Ordnance Survey of Ireland Discovery Series Map 27A, covering Cavan, Fermanagh, Leitrim and Monaghan (ISBN 978-1-901496-49-9)

Lough Oughter is the collective name given to a series of more than twenty interconnected lakes that are located to the north of County Cavan. Because of its patchwork profile, its overall area is hard to determine: winter expansion and summer reduction add to the problem. The lakes provide a delightful conduit for the River Erne as it meanders on its northward course. It is a wonderful venue for cycling, with very quiet country roads and lanes, allowing you to wander and explore the land squeezed between the lakeshores. You very rarely lose sight of the water for long periods, and the individual characteristics of the various lakes yield an ever-changing landscape that will capture your attention.

You will note that the distance for two circuits has been included at the head of this section. A road divides the northern section of the lake, and I did not think it feasible to adopt the same approach as with the intersecting road through Lough Gowna. The Lough Oughter road is 8 km long and instead of cycling a full circuit around all of the lakes, some cyclists may wish to choose this road, which still allows them to view most of the remaining northern lakes, albeit from a southerly perspective. This reduces the circuit to 35 km.

Another notable feature of Lough Oughter is the potential for numerous circuit extensions – more so than for any other lake in this book. I do not propose to cover all these extensions, instead allowing the reader to discover and explore them at their leisure. The cycle-route possibilities along the shores of this lake will provide the cyclist with many pleasant hours of both on- and off-road touring.

Killashandra

As the starting point for the circuit, I have chosen Killashandra, a pleasant village with a lake of its very own called, appropriately enough, 'Town Lough'. The village name is derived from the Irish cill a'sean ratha, meaning 'the church of the old ringfort'. This relates to a church that was built in the middle of an old fort that was sited beside the lake. There is a very small late-seventeenth-century church on the site, and the remnants of the old fort can still be seen. A planter family from Scotland called the Hamiltons developed the village in the early seventeenth century. Their fortified house, on the northern outskirts of the village, was destroyed by fire in 1911 – a fate which befell many of the houses of the gentry in the first quarter of the twentieth century.

Following the R201 northwards out of the village, you pass by the entrance gates of Castle Hamilton on the right and the Town Lough on your left. A small river running by the side of the road links this lake with Lough Oughter via Tullyguide Lough, which you will come to later on your right. Looking at the map, it is interesting to trace back in a south-westerly direction the route of this small river, sometimes called the Cullhies river, as it courses through a further seventeen small lakes, resembling a long necklace of pearls. A significant part of its course forms part of the county boundaries between Cavan and Leitrim and between Leitrim and Longford.

Killykeen Forest Park

On the right hand side, there is an opportunity to leave the R201 down a country road that runs along the shore of Tullyguide Lough. This road leads down to Killykeen Forest Park, a 243 hectare woodland with a diverse range of tree types, including oak, alder, beech, birch, sycamore and ash. The park is excellent for cycling where there are metalled roads, and has several trails along the lakeshore for diversions on foot. There is a pedestrian-footbridge link to the eastern side of the park. The route through Killykeen presents another opportunity for a short loop around the southern lakes of the Lough Oughter network.

Cloch Oughter Castle

Returning to the R201 for a short distance, the next turn right leads to the intersecting road, running across the northern part of the lake. Take this turn whether you are opting for the longer or the shorter loop. A left turn about 3.5 km along this road will take you up through the northern lakes. However, before you come to this, there is a turn to the right a short distance in from the R201 that presents the opportunity for a worthwhile extension. A 2 km-long lane by the side of Corglas Lough leads to the lakeshore, where there is a good view of the twelfth-century Cloch Oughter Castle, sited on an island near the shore. It is regarded as one of the best examples of the native Irish style of circular tower castles of the period from the thirteenth to the fifteenth century. The castle is sometimes referred to as Cloch Oughter Crannóg, as the castle was built on an island. Before the water levels of the lake were lowered, the walls of the tower were surrounded on all sides by water.

This Anglo-Norman fortress was once the principal seat of the O'Reilly clan. It was captured in 1224 from William Gorm de Lacy, who had built it a number of years earlier. In 1369, the castle fell temporarily to Philip Maguire, who was Lord of Fermanagh. Having regained possession, the O'Reillys continued in occupation until Sir Richard Wingfield secured control of the castle on behalf of King James I in 1607. Captain Hugh Culme was granted the castle and surrounding lands in 1610 as part of the plantation of Ulster, but the O'Reillys once again took possession during the 1641 Rising. Owen Roe O'Neill, one of the more prominent Irish chieftains of the seventeenth century, died in the castle in 1649. O'Neill was one of the leaders of an assembly of Catholics, known as the Confederation of Kilkenny, who came together to govern territory held following the 1641 Rising. Following the execution of Charles I in January 1649, Oliver Cromwell led a force to Ireland to recover the lands held by the Catholic Confederation. It took him a little over three years to complete this task. In 1653, Cloch Oughter Castle was one of the last strongholds in Ireland to fall to the Cromwellian forces.

If you wish to opt for the shorter circuit, you should continue east, where you will encounter two bridges some distance apart. The first is Carratraw Bridge, which crosses the River Erne as it passes between Carratraw Lough to the south and Inishmuck Lough to the north. The second bridge is Urney Bridge, which spans the Annalee river, a tributary of the Erne. This long river rises to the east in Lough Sillan at Shercock near the Monaghan border and is fortified by Dromore river as its passes near Cootehill. A sharp turn right after Urney Bridge will set you on a southerly course along the eastern side of Lough Oughter.

Drumlane round tower

Those opting for the longer loop should progress northwards by the shores of Inishmuck Lough, with its dominant island, which looks to be larger than the lake itself. Derrrybrick Lough lies to your left as you approach the tiny village of Milltown. Overlooking this lake on its western side are the ruins of Drumlane round tower and priory, which are sited on the grounds of what was an earlier, sixth-century monastery founded by St Mogue. Drumlane round tower is one of the less well known of the Irish round towers – and also one of the more unusual because of the sharp change in its masonry about halfway up. The lower half is smooth, and its construction is in keeping with similar structures dating back to the twelfth century. In contrast, the upper half appears to be of a cruder construction. Because the adjoining fifteenth-century church is of similar appearance to the upper half of the tower, one theory is that the additional brickwork was added to fit the tower to the church as a belfry. Another interesting feature of the tower is the engraving of a cockerel on the north face. The cockerel is believed to symbolise the resurrection.

After turning right at Milltown, there is a choice of routes across the northern extremities of Lough Oughter, depending on which way you wish to go around Drummany Lough. My preference is for the smaller country lane that trails along this lake's southern shore. Both routes meet up at Baker's Bridge, which spans the River Erne as it leaves the Lough Oughter system towards its northerly destination of Upper Lough Erne. The first turn right after the bridge leads across to the road that will link up with the route used in the shorter loop; from there, you will progress down the eastern shoreline.

A prominent landmark in the early stages of the journey south is rusty Derryheen Bridge, which spans the aforementioned Annalee river. There is a deep pool in the river at Derryheen known as 'the Wests', which is a popular fishing spot for anglers. After a sharp left turn at Derryheen, the peaceful country road leads past Lough Inchin into terrain that is a little removed from Lough Oughter but appealing nonetheless. A turn right at the next junction yields to a road that leads past several access points to the lake at Drumard and Inishconnell. The road also leads past the main entrance of Killykeen Forest Park on its eastern side. After this, you will lose sight of the lake for some time as you make your way south in the general direction of Crossdoney.

The circuit skirts north of this village, to link up with the R198/R199 just short of Bellahillan Bridge, which spans the River Erne. There is no option but to remain on the main road for the remainder of the journey to Killashandra. However, about halfway to the village there is a signposted turn to the right, leading down a 3 km stretch to Trinity Island, which features the ruins of a twelfth-century priory on its southern side. This is a pleasant extension that you can undertake before returning to Killashandra.

Lough Ramor

(25 kilometres)

LOCATION
Adjacent to Virginia off the N3, 81 km north of Dublin and 31 km south of Cavan town

LENGTH
5 km/3 miles

WIDTH
4.5 km/2.8 miles at its widest point

AREA
800 hectares/7.5 km^2

PUBLIC ACCESS
Foxes point at Virginia town on the eastern shoreline and Corronagh, almost directly opposite on the western side.

MAP
Ordnance Survey of Ireland Discovery Series Map 35 covering Cavan, Monaghan, Louth and Meath (ISBN 978-1-903974-25-4)

Lough Ramor is one of the smaller lakes included in this book, but this does not detract from its quality as a cycling circuit. It has an irregular shape: it is made up of a long, narrow southern section and a wide northern shore. Unlike a good number of the lakes covered in this book, Lough Ramor is not a limestone lake: it is located on older, more acidic bedrock, and as a result its waters have a darker hue.

Virginia

The starting point for the circuit is the car park at the shore access at Virginia, a couple of hundred metres off the main road. A ferry used to operate from this area to a point directly opposite on the western shore of the lake, to save people having to take a 10 kilometre trek by road. Virginia was founded as a plantation village in 1612 during the reign of James I and was named in honour of the Virgin Queen, Elizabeth I. It was a stop on the coach road between Dublin and Cavan and to this day enjoys a thriving passing trade from those journeying north and south. In 1750, the local landlord, the Marquis of Headfort, whose main house was on the outskirts of Kells, built a hunting lodge and sporting residence on the lakeshore north of the village. The building now houses the Park Hotel; in keeping with the sporting theme, a nine-hole golf course has been laid out in the 40 hectare estate attached to the hotel. The village also has a famous literary connection: Jonathan Swift wrote most of Gulliver's Travels while resident at nearby Cuilcagh House.

Deer Park Forest

Going northwards from Virginia, follow the R194 in the direction of Ballyjamesduff. The entrance to Deer Park Forest is about 2 km from the village. This wooded area offers walkers the opportunity of a 4 km-long trail that initially fringes the golf course and then skirts the lake on its north-eastern shore. The trail narrows as its leaves the lakeshore and runs northwards parallel to a stream before eventually emerging once again on to the Ballyjamesduff road, beside a stone cottage. For cyclists touring by road, at the first opportunity turn left off the R194 on to the Oldcastle road (the R195), which rises in a steady climb to provide good views across the wide northern expanse of the lake. Despite its small size, the lake's surface is studded with several islands.

The western shoreline

There is only one access point to the lake on its western shore. This is at Corronagh, almost directly opposite Virginia. The turn is signposted and the route down to the shoreline is a prolonged descent on a quiet country lane. Wood-framed chalets lie hidden behind trees in a private enclosure beside the shore. Fortunately, you will not have to retrace your tracks, as there is another lane to the left not too far along the return route that will lead you parallel to the shore for about 0.5 km. You will then need to prepare yourself for a steep climb away from the lake to return to the Oldcastle road. Before you reach that road, and while you are recovering from the ascent, take the time to absorb the best view of the lake, from beside the Catholic church, which sits on the brow of the hill just before you reach the main road.

Continuing south, you will probably be surprised to see two industrial facilities on the eastern shoreline. The largest of these is owned by Glanbia and used to be known as Virginia Milk Products. They are a blight upon the shorescape at this end of the lake, and one wonders how planning permission was awarded for them.

As you reach the southern extremity of the lake, there is a busy road that links the R195 with the N3 running parallel to the lakeshore. A low bridge spans the River Blackwater as its leaves its source to make its way through County Meath to Navan, where it feeds into the River Boyne. It is unfortunate that the last 6 km leg of this circuit has to be undertaken on a busy national route. However, the hard shoulder is of reasonable width and quality and affords some measure of safety from the speeding traffic almost all the way back to Virginia.

The Lakes of the

River Shannon

Lough Allen
Lough Ree
Lough Derg

The River Shannon is Ireland's longest river and one of its most important geographical features. With a total length of 334 km, it touches many parts of Ireland through its fortifying tributaries, which stretch for over 1,700 km, and meanders through fifteen lakes on its journey to the Atlantic Ocean. It drains about one-fifth of the entire island of Ireland. In earlier times, it was the artery through which Ireland's centre was explored. At one time, it was believed to have been a magical river that flowed two ways and had two mouths – a southerly one at Limerick and a northerly one at Ballyshannon. This was at a time before the discovery of the River Erne. Its fame has spread far afield: Ptolemy included it on his map of Ireland more than two thousand years ago.

The source of the River Shannon is a spring-fed pool located 52 m above sea level on the south-west slopes of the Cuilcagh Mountains near Glangelvin, County Cavan. The pool is known as the 'Shannon Pot'. Celtic mythology provides its own version of the river's source. According to legend, the river was named after the goddess Sinann, who in Celtic mythology was the daughter of Lodan, the son of Ler. Ler was one of the more prominent figures in Celtic legend, being the sea-lord of the people of Dann, the Tuatha Dé Danann. Legend has it that Sinann went in search of forbidden knowledge at a sacred well. On her approach, the well erupted and swept her away in a flood of water, creating the waterway over which she now proudly presides.

The banks of the river have been the scene of much bloodshed through the ages. In the days of early Christianity, monastic settlements were established nearby, and the principal islands proved popular retreats for holy men. Foreign invaders, such as the Vikings and later the Normans, used the river to access native strongholds and in time established their own bases along the river. The Vikings destroyed some of the monasteries, notably those at Clonmacnoise and Clonfert. In more recent centuries, the river served as the focus of a water-based transportation network, with man-made links to Dublin via the Grand and Royal Canals. At one time, for a very brief period, the Shannon was also part of a navigation network that linked Belfast to Limerick. Faced with unyielding competition from roads and railways, 160 years of commercial transportation on the River Shannon finally ended on 10 June 1959, when Barge 41M made a delivery of twenty tons of flour from Limerick to the quayside at Carrick-on-Shannon in County Leitrim.

Nowadays, the River Shannon and its lakes are mostly associated with a range of recreational activities, including cruising, sailing and angling. The use of the river for recreation is not a new phenomenon, however: there is a record of a regatta having taken place at Athlone as far back as 1731. The Lough Ree Yacht Club is said to have been established in 1770, making it the second-oldest yacht club in the world, while the Lough Derg Yacht Club, founded in 1836, has also stood the test of time. Navigation links with Dublin exist via the Grand Canal, and a second link via the Royal Canal will be re-established as soon as the ongoing restoration of that waterway is complete. Another important link was re-established in 1994, with the restoration of the Ballinamore–Ballyconnell Canal (now known as the Shannon–Erne Waterway), a blend of still-water canal, lakes and canalised river that links the mighty River Shannon with the vigorous River Erne. These linkages are vital to maintaining the Shannon as a vibrant river, but it is the peace and tranquillity to be found in many places along its banks that will remain its most endearing attraction.

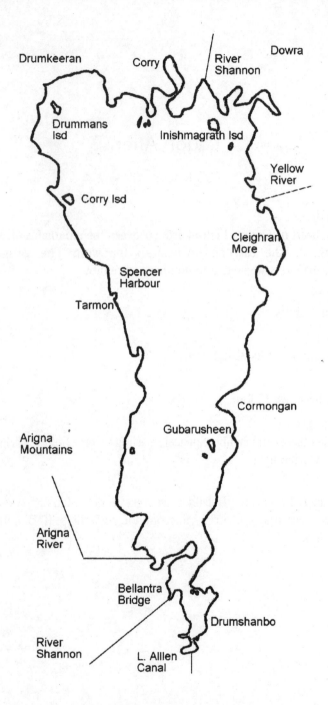

Drumkeeran

Corry

River Shannon

Dowra

Drummans Isd

Inishmagrath Isd

Yellow River

Corry Isd

Cleighran More

Spencer Harbour

Tarmon

Cormongan

Gubarusheen

Arigna Mountains

Arigna River

Bellantra Bridge

Drumshanbo

River Shannon

L. Alllen Canal

Lough Allen

(53 kilometres)

LOCATION
In the heart of County Leitrim, lying between the mountains of Slieve Anierin to the east and Arigna to the west. The village of Drumshanbo is situated at its southern extremity.

LENGTH
13 km/8 miles

WIDTH
5 km/3 miles at its widest point

AREA
3,500 hectares/35 km²

PUBLIC ACCESS
Wynne's bay Drumshanbo, Spencer's harbour, Corry Strand, Cleighran More, Cormongan

MAP
Ordnance Survey of Ireland Discovery Series Map 26, covering Cavan, Fermanagh, Leitrim, Roscommon and Sligo (ISBN 978-1-901496-48-2)

Lough Allen is the first of three great lakes encountered as the River Shannon makes its descent to the sea. Long and narrow, it is sandwiched between the Arigna Mountains to the west and Slieve Anierin to the east and divides the county of Leitrim. It will surprise many to learn that it used to act as a reservoir to maintain water supplies for the Ardnacrusha Hydroelectric Power Station, located at Killaloe, which is considerably further south on the River Shanon near Limerick. Sluice gates at Bellantra near Drumshanbo were used to control the flow; when these were operating, the water levels of the lake used to fluctuate quite a bit, making the lake dangerous for navigation at low water levels. The lake has only recently joined the Shannon Navigation system on a full-time basis, and new mooring facilities have been installed at Spencer harbour on the north-western shore and Cleighran More on the eastern shore. Another danger for water-borne users of the lake is the strong winds which can rise suddenly, funnelling down between the mountainous terrain along the lake's shores. Encounters with the lake's mythical monster, 'massive with several bumps and an enormous head', might also prove entertaining to boat users.

Despite being surrounded by mountains, the cycling route around Lough Allen is relatively flat, apart from a steady climb into Drumkeeran and a roller-coaster ride along the country lanes leading away from that village. The winds can present a challenge at times, particularly on the eastern side, under the lee of Slieve Anierin, where the road is exposed. Most of the route is on main roads, but this should not be too offputting, as the roads in this area are generally quiet, even in summer. Apart from a section along the northern end of the lake, where you have to divert to Drumkeeran, for much of the journey you will be able to remain reasonably close to the shores of the lake and benefit from some delightful views.

Drumshanbo

The starting point for the circuit is the pleasant shore-access facility at Wynne's bay, on the westerly outskirts of Drumshanbo, close to the junction with the Lough Allen Canal. The village is located on the southern tip of the lake and has good refreshment facilities. Its name, derived from the Irish for 'Bridge of the Old Hut', is thought to relate to a cave dwelling that existed along the shores of Lough Allen in prehistoric times. It is a pretty village, and one of its attractions is the

Slieve Anierin Visitor Centre, which relates the history of the area, with particular focus on mining, lakes and the waterways. The centre includes the reconstruction of a sweathouse, a common feature of the locality. Constructed of stone, the ancient sweathouse had the same purpose as the modern dry-air sauna but had to be heated by fire for a whole day before it was needed.

Lough Allen Canal

The Lough Allen Canal, built in 1822, links the lake with the Shannon–Erne Waterway at Leitrim village. In recent years, a new lock was installed at Drumshanbo: this lock is unique in that it provides a two-way system to allow for fluctuating water levels in the lake.

The Miner's Way

The initial journey west from Drumshanbo is very encouraging for cyclists. Indicated on the map is a route that follows a waymarked trail called 'The Miner's Way': this looks like a good route to track for the initial stages of the loop, as it stays close to the lake shore. The Miner's Way traces the paths that coal miners used to follow on their way to work in the coal-mining territory of the Arigna Mountains. A turn right where it is signposted for Holly Island leads quickly on to country lanes, which bring you to Bellantra Bridge, an ugly-looking iron structure with sluice gates at the bottom and a winding mechanism at the top. After you have crossed this, you will come on to an even narrower country trail, which is well surfaced – and well used by cattle. (The biggest problem you will face is avoiding the cow-pats.) So far, so good! Arriving at a junction, the waymarked-trail signpost points down another small lane, with grass running down the middle. This is the time to resist further pursuit of the trail: you should continue straight on, to return to the R280.

For those of you who are curious as to what would happen if you continued down that appealing boreen, let me enlighten you. The early promise evaporates, as the trail leads to a farmyard, and there is no immediate indication as to which way one should continue. Looking around, you will notice a small pedestrian gate to the right of the farm gate, with a rusty weight attached to it to ensure that it swings back. This leads into a field, which ends in an extremely muddy and boggy path between gorse bushes and brambles. The path is not suitable for cycling and is barely suitable for walking. Knee-deep in mud and cow dung, you emerge into a small clearing, where, on several visits I made, there were no signs indicating the route – yet by reference to the map you know you are in the right place. By crossing three barbed-wire fences and a deep, muddy trench, you eventually discover a sign indicating that the path continues to the left of a small winding river, the Arigna river. In case the trail was not difficult enough, you have to haul your bike with you. The path can get completely overgrown, and you may have to tunnel your way through the vegetation: very apt, considering the name of the route! Emerging into another small clearing, the signage indicates that the way continues on a path that is divided from a field by barbed-wire fencing. Unfortunately, the path was not passable at the times I visited, and I am sure that the local landowner would not be enamoured with people climbing over the fence and trespassing on his field in order to make further progress. The trail eventually emerges on to the R280 at a stile close to Mountallen Bridge, which spans the Arigna river.

Arigna mines

If you look to your left just after the junction with the R285, you will see an ivy-clad stone bridge spanning the Arigna river. This bridge, which dates from the nineteenth century, was built as part of the engineering works associated with another canal linking the river to the nearby Shannon; this canal was proposed for the purpose of transporting coal from the mountains for shipping out using the Shannon. The government had been persuaded that a coal seam in the area had been measured at 5.5 m, whereas in reality it only reached down to a depth of 46 cm. The canal builders were jailed for fraud, having been given £800,000 to help finance their enterprise.

Mining in the Arigna Mountains is believed to have started as early as the fifteenth century, when iron was initially extracted. Coal mining predominated from the nineteenth century until as recently as July 1990, when the last of the mines were closed down. At the time of closure, there were about six or seven commercial mines operating. Guided tours illustrating the working of the mines and the conditions endured by the miners are now available.

Tarmon

The road rises gently as you pass through the townland of Tarmon, yielding good views across the lake. Close to the shore are about half a dozen nursery beds for young salmon, which are nurtured here in freshwater before being introduced to the sea. The ruins of St Patrick's Abbey can be seen to the right, close to the lakeshore. A one-time resident of this area was a traditional flute player, John McKenna, who was born on the mountainside, at Tents, in 1880. McKenna emigrated to New York around 1911 and died there in 1947. He made several recordings of his music while in the United States and is commemorated in a memorial built on the roadside, which you will see opposite the local parish church at Tarmon.

Spencer harbour

Just beyond Tarmon is a Waterways Ireland Marina called Spencer harbour. The marina is located beside an old brickworks with a red-bricked chimney, and the derelict ruins of buildings including Spencer House. The site was once the home and workplace of more than two hundred people. The brickworks, the house, the private harbour and the nearby five-acre Corry Island were offered for sale in April 2006 for the princely sum of €600,000. Spencer harbour used to be an important port of call on the Royal Canal system and was used to ship coal from Arigna Mines and cast iron from Creevelea Ironworks. It is said that part of Dublin's Ha'penny Bridge was cast at those ironworks. While I am aware that navigation on Lough Allen has only recently become a full-time part of the Shannon Navigation system, one has to wonder about the value for money offered by this facility: the area is isolated, with no shops or other services nearby.

Tour de Humbert Cycling Trail

As you embark on a long, steady climb towards Drumkeeran, you will encounter a signpost indicating the route for the Tour de Humbert Cycling Trail. This tour follows the route of the march taken by the French general Jean Joseph Amable Humbert at the head of a revolutionary expedition to Ireland during the 1798 Rebellion. It extends 225 km from Mayo and Sligo, through Leitrim, and on to Ballinamuck in County Longford, where the rebels were defeated.

Drumkeeran

Skirting the quiet village of Drumkeeran, it is interesting to reflect on the impact that the Great Famine of the 1840s, and subsequent emigration, had on remote villages such as this. In 1841, before the Famine, the population of the village was recorded as being 9,491; today it stands at around 1,000. The same trend is repeated throughout the whole of Leitrim: between 1841 and 1851, the county's population dropped from 155,000 to 112,000, and today it stands at 25,000. The poor agricultural quality of the land, and the lack of other indigenous industry, paved the way for sustained high emigration, resulting in the massive decline in population. Evidence of this decline is to be seen everywhere, with the ruins of abandoned cottages around the shores of Lough Allen and other lakes of the region being a frequent sight. In recent years, new residences have sprung up, but a good many of these are holiday homes or weekend retreats and thus are doing nothing to swell the indigenous population.

Leaving Drumkeeran, you get the opportunity to avail of country lanes for several kilometres. Turn right as you enter the village, and use the small road facing the church. This leads down a roller-coaster ride, with speedy descents and slow ascents coming in rapid succession for over 3 km, until you join the R200, which will lead you towards Dowra. Along this road there is a signposted turn leading to Corry Strand, a pleasant picnic area with a small jetty and good views across to Drummans Island and Corry Mountain, where wind-farm turbines now dominate the skyline. This is the only authorised access area on the northern shore of Lough Allen.

Dowra and the eastern shoreline

Returning to the R200, the route to Dowra takes you away from Lough Allen, and you cross the county boundary with Cavan just north of the Owennayle river. Dowra, the first town on the River Shannon, is on the boundary line between counties Cavan and Leitrim. In September 2006, Waterways Ireland announced that it had plans to extend the navigable channel from Lough Allen to Annagh Upper, which is located in County Leitrim, approximately 3.5 km from Lough Allen and less than 2 km downstream from Dowra village. Work was due to be completed by 2008. The new works are to include a lay-by to accommodate a sixteen-berth public marina, consisting of floating walkways and small jetties at which boats may be moored.

The north road out of the village leads to Glangelvin, which is close to the Shannon Pot, the source of the River Shannon. Leaving Dowra, the journey south along the eastern side of the lake begins on the R207, but at the first opportunity as you exit the village, you should turn right and join the route of the Kingfisher Cycle Trail for a short while. You will encounter signposts for this trail frequently in this region as it loops around the counties of Fermanagh and Leitrim, covering more than 370 km in all. The area to the right, between the country lane you are cycling on and the banks of the River Shannon as it makes its way to the lake, forms part of what was known as the Black Pig's Dyke – the name given in earlier times to the frontier between the province of Ulster and the rest of Ireland. (See the section on Upper Lough Macnean, pp 154-58).

Rejoining the R207, continue south through the small village of Ballinagleragh, and onwards past the Lough Allen Adventure Centre. A brief diversion can be made to visit St Hugh's Holy Well, or 'Tobar Bheo Aoidh', which is signposted to the left just after the Adventure Centre. There is a small picnic area beside the well, where you can enjoy the views across the lake. There used to be a pilgrimage to this well on the saint's feast day, 8 March, but this no longer happens. Near the well, you will get the chance to inspect closely one of the many sweathouses in the area.

A new marina facility has been built recently in the small bay at Cleighran More, which is protected, by a promontory, from the winds that can suddenly whip up along this exposed shoreline.

The last shore access on the eastern side before Drumshanbo is located a couple of kilometres further south, at Cormongan. From this point on, the road rises gently, and as you approach Drumshanbo you can enjoy some excellent views across the lake before returning to Wynne's bay.

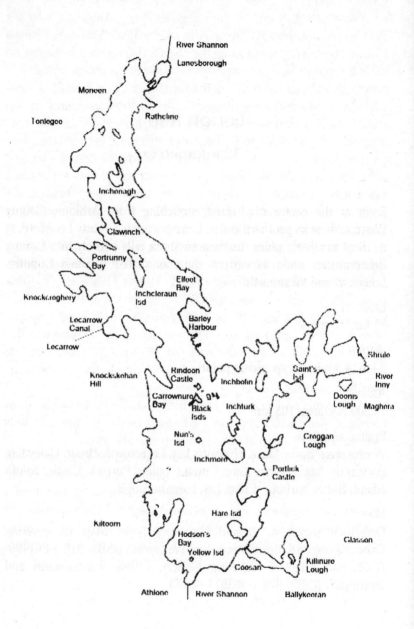

River Shannon

Lanesborough

Moneen

Tonlegee

Ratheline

Inchenagh

Clawinch

Portrunny
Bay

Elfoot
Bay

Knockcroghery

Inchcleraun
Isd

Lecarrow
Canal

Barley
Harbour

Lecarrow

Shrule

Knockskehan
Hill

Rindoon
Castle

Saint's
Isd

River
Inny

Inchbofin

Carrownure
Bay

Doonis
Lough

Maghera

Black
Isds

Inchturk

Nun's
Isd

Creggan
Lough

Inchmore

Kiltoom

Portlick
Castle

Hare Isd

Hodson's
Bay

Glasson

Yellow Isd

Killinure
Lough

Coosan

Athlone

River Shannon

Ballykeeran

Lough Ree

(120 kilometres)

LOCATION
Right at the centre of Ireland, stretching from Athlone, County Westmeath, at its southern end to Lanesborough, County Longford, at its most northerly point. Its western shore falls entirely into County Roscommon, while its eastern shore is shared between Counties Longford and Westmeath.

LENGTH
25 km/15 miles

WIDTH
7 km/4.3 miles at its widest point

AREA
10,500 hectares/105 km^2

PUBLIC ACCESS
Western shore: Barry More, Hodson's bay, Lecarrow harbour, Galey bay, Portrunny bay. Eastern shore: Coosan point, Portlick Castle, Saint's Island, Barley harbour, Elfeet bay, Lanesborough.

MAP
Ordnance Survey of Ireland Discovery Series Map 40, covering Galway, Longford, Roscommon and Westmeath (ISBN 978-1-901496-25-3), and Map 47, covering Galway, Offaly, Roscommon and Westmeath (ISBN 978-1-903974 –23-0)

Lough Ree is the second largest of the Shannon lakes and is perhaps less impressive than its larger and smaller sister lakes, Lough Derg and Lough Allen. Perhaps this is because it lacks those two lakes' dramatic backdrop of rugged hills and mountains, as it is situated in the midst of the flat plains of the Irish midlands. There are few significant areas of rising ground from which to observe and appreciate its broad expanse, its numerous bays and inlets, and its many islands. From a cyclist's perspective, its flat hinterland means that for long stretches the lake, while nearby, remains out of view. To many cyclists, that would seem advantageous, as it offers a relatively flat course and allows a speedy circuit of the lake, but the lake is probably better appreciated from the deck of a boat rather from than the saddle of a bicycle. It is not all bad news for the cyclist, however. The circuit around the lake will introduce you to areas of hidden Ireland that are relatively untouched by tourism and allow you to venture through minor roads and country lanes bounded by a varied landscape of bogland, farmland, woodland and wetlands. For significant parts of the circuit traffic will not be an issue, so from that viewpoint alone this is a journey well worth undertaking.

General description

Lough Ree derives its name from the Irish 'Loch Ridh', or 'Lake of the Kings'. Often described as an inland sea, it is a generally shallow limestone lake, although depths of 35 m have been recorded in some parts. Its many islands tend to be larger than those found on the grander Lough Derg. Its northern half is narrow, with an average width of 3 km, while the southern half is considerably broader, with an average width of 11 km. In the south-east corner, there are a number of smaller connected lakes, collectively known as the 'inner lakes'. These are Coosan Lough, Killinure Lough and Ballaghkeeran bay.

The Lough Ree Trail

The Lough Ree Trail is a driving route centred around the lake that is sometimes half-heartedly marketed as a cycling route. Unfortunately, it concentrates on national roads and extends too far away from the lake to be considered to be a genuine cycling route. I hope that the circuit outlined in this book will go some way towards redressing this situation.

Athlone

Athlone is an appropriate starting point for the circuit, as it occupies a strategic crossing point on the River Shannon at the base of Lough Ree. George Petrie described it in the 1820s as 'a sorry Irish town, with little to boast of, except the noble river on whose banks it is placed'. The town's name is an anglicised version of the Irish 'Ath Luain', or 'the Ford of Luain'. The annals suggest that Luain was an ancient innkeeper who presided over a hostel at the crossing point and that he used to guide travellers across the river by means of stepping stones. Because of its pivotal position both at the centre of Ireland and at the narrowing of the Middle Shannon, it is understandable that the town became a focus for various communities. The first bridge across the river was said to have been built in 1129 by Turlough O'Connor, king of Connacht, who also built Athlone's first castle, no doubt to protect his territories in the west from marauding Normans. He was evidently quite successful in this, because it took the Normans nearly sixty-five years to conquer Connacht, after quickly overrunning the eastern part of the country. When they did eventually cross the Shannon, they built more enduring stone castles than the wooden structures erected by Turlough O'Connor. The imposing castle now sited on Athlone's left bank was built in 1210 by one such Anglo-Norman, John de Grey, Bishop of Norwich and Irish Justiciar to King John. Today, the castle is known as King John's Castle, and it houses a museum, a visitors centre and a tourist office.

Religious communities were also drawn to Athlone, with monasteries being established on both banks. A little-known order called the Cluniacs established their only Irish monastic settlement on the west bank in the thirteenth century, and while their physical connection with the town has long since disappeared, their enduring legacy was the introduction of eel fishing to the locality – an activity that is now carried on by local fishermen on behalf of the Electricity Supply Board. The Franciscans built their monastery on the east bank, and their connection with the town survives, with a Catholic church located in the eastern part of the town.

The bridge at Athlone has been through several incarnations since Turlough O'Connor's original wooden structure. In 1566, Sir Henry Sidney spanned the river with the first stone bridge, and in 1691 this bridge was the scene of one of the fiercest battles to be fought on

Irish soil. Williamite troops, fresh from their victory at the Battle of the Boyne, were repulsed by the governor, Colonel Richard Grace, and his soldiers, only to return the following year in a fresh attempt to take the town – in a battle that has gone down in history as the Siege of Athlone. During the battle, a non-commissioned officer called Sergeant Custume led a brave attempt to blow up the bridge in order to prevent the Williamites from crossing the river. The attempt failed, and he was killed. During the battle, it was said that as many as 21,000 cannon balls were used to knock down the castle, which was subsequently rebuilt. The barracks in the town is named in his honour: the only military post in the British Isles to be named after a non-commissioned officer. A new bridge was built in 1844, and this has survived to the present day.

Hodson's bay

Opting for a clockwise circuit, the initial stages require two early detours in order to get close to the most southerly part of Lough Ree. At the first opportunity, turn right, leading to Barry More, where you will get a good view of where the lake narrows, to emerge once again as the River Shannon. After backtracking to the N61, there is another detour past Athlone Golf Club to Hodson's bay, where there is a harbour and a fine hotel, which used to be the family home of the Lenihan family, which has provided prominent members of the Irish government across two generations: in the late twentieth century, Brian Lenihan senior and his sister Mary O'Rourke both held senior Cabinet posts, while Brian Lenihan junior and his brother Conor have been minister and minister of state, respectively, in more recent administrations. To the north of the harbour is Hodson's Pillar, said to be the geographical centre of the island of Ireland. Backtracking once again to the N61, you will be without a view of the lake until you finally get a chance to turn off on to minor roads just beyond Kiltoom, following the signpost for Rindoon Castle and Caragh. The journey to the castle is through hedge-lined country roads and over Knockskehan Hill, where at last you get a decent view of the lake looking across Carrownure bay to the Rindoon peninsula.

Rindoon Castle

The ruins of Rindoon Castle lie at the end of the peninsula. The name Rindoon, sometimes given as 'Rindown', means a fortified headland. In order to access the ruins, you have to go down a laneway that leads to a farmhouse. This is one of those areas where you are unsure as to whether you can progress further, as farm gates block your way. There is no indication in the available guidebooks as to whether rights of way exist, nor does the panel at the entrance warn that you may be entering private ground. You will have to make up your own mind. You will, however, note that there are signs in the area warning of bulls. It would appear that landowners in this locality employ guard bulls rather than guard dogs to protect their property! Despite the warning of the presence of bulls, I have ventured more than once up to the site. While enjoying the view across the broadest part of the lake, it is easy to understood why the castle was located on this peninsula: the castle commands views both directly across the lake and also northwards and southwards almost the full length of the lake.

The ivy-clad castle ruins are those of a Norman fortification which was built on the site of an earlier Viking settlement. It is in fact one of the few royal castles built in Ireland: that is, it was constructed on royal command rather than by one of the great lords. The original castle was built between 1227 and 1235 by Geoffrey de Marisco and was in royal hands for much of the thirteenth and fourteenth centuries, during which time it was subjected to many attacks, including being levelled in 1272. It was frequently repaired and enlarged during this period, and different phases of building can be observed among the ruins, although outer defensive installations such as wooden towers and a drawbridge have long since disappeared. There are other historical features on the peninsula, including a ruined church and the remains of a wall that once surrounded a town that was sited here in medieval times.

Lecarrow Canal

There is a small loop on tight country lanes that you can include as you leave the entrance to Rindoon Castle, allowing you to skirt St John's Wood. The wood is described on the information panel at Rindoon as being unique, in that it has grown on shallow soil over a bedrock of

limestone (which is difficult for roots to dig into). The loop leads back
to the connecting road to Lecarrow. Along this road there are fine
views across Blackbrink bay towards Inchcleraun, the second-largest
island on the lake – about which more will be said later. Lecarrow is a
small village located just off the N61. There is a small tree-lined harbour located at the end of a canal that leads down to the lake. The 2
km-long canal was constructed in two stages in the 1840s. The section
nearest the lake was carved out in 1841 in order to facilitate the transport of stone from Lecarrow quarries to Athlone, where it was used
in the building of the bridge and quayside; the extension, to the village, was cut later, as part of a famine-relief programme. It later fell
into disuse and became hidden under heavy growth until the 1960s,
after which time it was reclaimed, with the support of the Inland
Waterways Association of Ireland. It now comes under the auspices of
Waterways Ireland.

Knockcroghery

From Lecarrow, you will have to resort to the N61 for the last time,
heading in the direction of the next village, Knockcroghery, famous
for the manufacture of clay pipes for smoking. The village name has
unusual origins. The Irish version is 'Cnoc an Chrocaire', meaning
'Hangman's Hill', which refers to the hill near the village which was
once used as a place of execution. Before you reach the village, branch
off to the right before the railway junction, to avail once more of
country roads that will allow you to get closer to the lakeshore. At the
next crossroads, a turn right leads down to Galey bay, where there is a
caravan and camping park, and the ruins of a fourteenth-century castle which was once occupied by a local chieftain, William Boy O'Kelly.
From the crossroads, the road rises, passing the small lake called
Lough Oura. When you reach the top of the hill, you will get the first
view of the tall chimneys of Lanesborough power station – a useful
marker for the rest of your journey north. Turning right at the bottom
of the hill will bring you to Portrunny bay, a popular boating stop with
a small harbour.

Inchcleraun Island

From Portrunny bay, there is a good view of Inchcleraun Island. The
island is named after Clothra, who is said to have established a centre

of learning on the island. She was a sister of Queen Maeve of Connacht, a formidable warrior goddess in Celtic mythology mostly associated with the legend of the Táin Bó Cuailnge (The Cattle Raid of Cooley). While bathing on the shore of the island, Maeve met her end after being struck by a sling shot fired by the son of the king of Ulster, whom she had fought against while capturing the Brown Bull of Cooley.

The ruins of a monastic settlement established by St Diarmuid in 540 are to be found on the island, whose alternative name, Quaker Island, is attributed to a previous owner, a Mr Fairbrother, who was a Quaker.

Green Heartlands Cycle Route

Backtracking slightly from the harbour at Portrunny bay, a small road to your right allows you to track the lakeshore through one of the most appealing cycling sections in the circuit. The low-lying nature of the circuit means that you get only limited views of the lake, however. By way of compensation, the route takes you through bogland and woodland where you are unlikely to encounter a car. As you roll along country lanes, you will come across signposts for the Green Heartlands Cycle Route. This is a continuous 217 km-long circuit, comprising northern and southern loops around south Roscommon, that was designed and laid out along quiet country roads in the late 1990s. Unfortunately, it was promoted only on a restricted basis, and some of the markers along the route have now disappeared. You will have to pay close attention to the map to ensure that you do not stray on to the tracks that lead through the bogs themselves. Eventually, you will emerge on to the N63 at Tonlegee junction, which lies approximately midway between Roscommon town and Lanesborough. Before you reach Lanesborough, you can get off the busy national road for a couple of kilometres and once more enjoy nice views of the lake: turn right at Moneen, and this short loop will lead you back to the main road just short of Lanesborough.

Lanesborough

When you arrive at Lanesborough, you are at the very head of Lough Ree. You have crossed over from County Roscommon into County

Longford and from Connacht to Leinster, with the River Shannon providing both the provincial and county border. The town owes its name to Lord George Lane, who was granted the now-ruined castle and surrounding lands by King Charles II in the seventeenth century. It was previously known as 'Ath Liag', which translates as 'the Ford of the Flagstone'. The smokestacks of the ESB power station dominate the town. Hot water is pumped from the power station into the river, and as a result the semi-tropical environment harbours unusual varieties of fish and makes the area a popular place for anglers. There are three distinct points of shore access in Lanesborough: the harbour to the east of the town, the pier to the north of the bridge, and a slipway to the south of the bridge.

Rathcline

As you leave Lanesborough behind, taking the right turn at the junction in the centre of the town, you once again lose sight of the lake for a fairly long period. About 4 km from the town, you can make a small detour to view Rathcline Church, whose cemetery features the ruins of an ancient church dating back to the twelfth or thirteenth century. Look out for the Síle na Gig carving on the northern splay of the window at the high end of the west gable. Such carvings, which are etched into the stonework on religious buildings in Ireland and elsewhere, are thought to have pagan origins.

Returning from the Rathcline detour, resume the southerly course with care. This is an area that requires repeated reference to the map to ensure you do not end up in too many dead ends, however delightful the views they provide over the islands of the northern part of the lake. Your target should be the Cashel area on the shores of Elfeet bay, where there is a ruined church with a cemetery still in use. If you continue down the hill past the church, you will come to a small harbour, which is used to transport livestock to and from nearby Inchcleraun. Retracing your steps, turn sharply right at Cashel church, where the road rises through a small area of dense woodland and emerges at a parking bay, which provides the best views of the narrow northern part of the lake. Underneath, to the left, lie the ruins of a fortified tower house that was once the home of George Calvert, who was governor of Maryland and founder of the city of Baltimore.

Barley harbour

Descending from the viewing area, take the second turn right, and then right again, to continue through deciduous woodland in the direction of Barley harbour. This is the only deep-water harbour in County Longford, on the eastern shore, and was built in the 1960s. Before you reach the harbour, look out for an interesting holy well hidden away down a grassy lane. Known as 'Lady Well', this has drawn people seeking cures since the mid-1800s, and its water is gin-clear and cold. As is the case with other holy wells you will encounter on your travels, there is a tree beside the well festooned with rags and surrounded with religious items.

Saint's Island

Following the directions for Saint's Island, you are tracking the lake's eastward extension. It is worthwhile making a detour to Saint's Island, which, despite its name, is not an island but a small promontory stretching out into Inny bay. Here you will find the ruins of St Catherine's Church and burial ground. Looking out into the bay, you will see Inishbofin Island, meaning 'the Island of the White Cow'. On the northern part of the island are the ruins of two churches close together; the earliest one was built by St Rioch, a contemporary of St Patrick.

Backtracking from Saint's Island, rejoin the small country road that leads eastward through the townland of Forthill, after which you come to a crossroads, where you turn right in the direction of Ballymahon. This road brings you away from the eastern extremity of Lough Ree but offers a very pleasant cycling environment. Straight, flat, well-surfaced roads cut through actively worked bogland, and with little traffic in the area, the ambience is extremely tranquil. You would imagine that, with the flat landscape, there would be views of the lake, but in fact the lake is so low-lying that it is completely out of view. In addition, sections of the road are bounded by high hedgerows – which contribute to the quiet but also restrict the view.

'Leo'

After approximately 3 km, the road turns sharply right, setting you on a southerly course towards the townland of Shrule, through which the River Inny passes on its way to the lake. In this area, you will come across a number of signposts simply bearing the name 'Leo', with a plume alongside, pointing to a particular landmark such as a school or graveyard. When you come to the bridge at Shrule, you will understand the relevance of these signs: Leo was a poet and patriot called John Keegan Casey (1846–70) who hailed from this area and worked as a clerk in Shrule Mill, on the banks of the Inny. In poetry terms, he is overshadowed in the area by the more celebrated Oliver Goldsmith, who was born 6 km to the west in Pallas – and about whom more will be said later. Casey was arrested as a Fenian and was imprisoned and later executed in Mountjoy Jail in Dublin. There is a memorial to the poet near the bridge, where you can sit on a bench, sheltered by a tree, while you read verses from his poems, which adorn the walls leading up to the bridge. Reminiscent of the war poet Francis Ledwidge, who hailed from County Meath and also died young, Casey drew inspiration for his poetry from the river and countryside that gave so much pleasure to him during his youth:

> The Inny
>
> No ship cast anchor on its soft bright tide
> But always silenced by its flowery brink
> Save when a dreamer sought its quiet side
> And there on God's bright beauty sat to think.

Sadly, Leo did not fulfil the wish expressed in the following lines:

> Oh I would die beside a lonely river
> Whose waves should pour a flood of eloquence upon my ear.

Murray's Old Style Pub

Leaving Shrule with the words of Leo echoing in your head, you emerge on to the busy N55 for a brief cycle to the unusually named village of Tang, where you can resume on minor roads by turning right in the direction of Maghera. Along the quiet road to Maghera, you will come across an isolated but welcome oasis in the form of Murray's

Old Style Pub, which certainly lives up to its description. The entrance to the pub, which is housed within a converted cottage, challenges anybody of moderate height. Its interior is dark and intimate, and when a stranger enters, conversation stops momentarily as its inhabitants gauge whether the interloper is friend or foe. Perhaps it is the sight of a backlit, windswept cyclist in shorts that is the conversation-stopper. Within moments, though, you will find yourself immersed in light-hearted chatter, satisfying the locals' curiosity in relation to your cycling endeavours. I found out later from the proprietors of the Village Tea Room in nearby Glassan that, despite its remote location, Murray's is an extremely popular venue for traditional-music sessions during the week.

Portlick bay and castle

At this stage, you will have been without sight of the lake for a considerable number of kilometres. In order to address this situation, turn right at the townland of Lackan, where the road from Maghera comes to a dead end. This leads down to Coolaleena crossroads, where several of the meeting roads could once again lead you to views of the lake. A turn left sets you on course for Portlick, the only official shore-access point, which looks out directly on to Lough Ree's largest island, appropriately called Inishmore or Inchmore. Portlick Castle, to which there is no public access, dates back to the fourteenth century and was built for a Norman family called Dillon. It is said to be one of the oldest surviving Norman castles. There is a delightful off-road trail along the lakeshore at Portlick bay. The trail passes through woodland that was planted in 2000 as part of the People's Forest Programme to celebrate the millennium.

Hare Island and the inner lakes

From Portlick, you can maintain close contact with the lakeshore through the townland of Killinure and can venture towards Killinure point, although there is no public access up to the point. However, the public-access road will give you good views of Hare Island, where, in 1802, the largest horde of Viking gold ever to be found in Europe, dating back to the tenth century, was found. The island used to be called Inis Aingin: well before the arrival of the Vikings, St Ciaran, who later founded the more celebrated Clonmacnoise monastery, established a monastic settlement here. Despite his famed association with Clonmacnoise, St Ciaran died only seven months after establishing it, having spent a significantly longer period at his retreat on Hope Island. Killinure point marks the narrow entrance off Lough Ree to the inner lakes of Killinure Lough, Coosan Lough and Ballaghkeeran bay, with Coosan point opposite. Squeezed between Coosan Lough and Ballaghkeeran bay lies Friar's Island, to where the Franciscans moved in the seventeenth century, having earlier maintained a monastery on the east bank of the river at Athlone. In one guide for the area, I read that the Annals of the Four Masters are said to have been written at the Franciscan monastery here in 1628. This is unlikely, however, as it is widely accepted that they were compiled much later in the seventeenth century, between 1682 and 1686, by a Franciscan Brother and three lay associates in a house of the Franciscan friars of Donegal by the banks of the River Drowes, which flows into Lough Melvin (see page 148).

Glassan and Oliver Goldsmith

You have no alternative but to turn inland again, towards Glassan, to continue the circuit. This is no hardship, however: it is a pleasant village that takes its name from the Irish word glasán (small stream). It is said locally that this is the village of 'Sweet Auburn' that features in the opening line of the poem 'The Deserted Village', composed by Oliver Goldsmith (1728–74). Goldsmith, one of Ireland's most celebrated poets, novelists and playwrights, was said to have been born in Pallas, close to Ballymahon. If you were so inclined, Pallas could be incorporated into your lake circuit by means of a relatively short diversion west from the village of Tang, which you will have passed through earlier. Goldsmith was born on 10 November 1728, the son of a clergyman, and was a student of Trinity College Dublin from the early age of fourteen. He is best remembered for the aforementioned poem, published in 1770, and describes life in his native countryside, a portion of which you will have just experienced. It has been said that Goldsmith may not have been born in Pallas but instead at the home of his maternal grandmother near Elphin, County Roscommon, on the other side of the River Shannon: he was born prematurely while his mother was visiting her. In any event, there is a statue at Pallas marking the location of the Goldsmith homestead. The Oliver

Goldsmith Summer School takes place every June and July in Abbeyshrule and Ballymahon.

Glassan, known as 'the Village of the Roses', was built in the 1740s as an estate village for the labourers of Waterston estate. A pleasant stop-off point is the Village Tearoom, located in the Glassan Arts & Crafts Centre, where you can avail of refreshment while enjoying the display of creative talent of the resident artist.

It was near Glassan that Lough Ree's renowned monster was most recently spotted. Tales of a large water creature lurking in the depths of this lake have been passed down through the generations. In his book Irish Ghost Stories, Patrick Byrne quotes the following passage from 'The Life of Saint Mochu', a sixth-century saint: 'the lake is infested by a monster which is accustomed to seize and devour swimmers'. The last reported sighting was in 1960, by three clerics on a fishing expedition near Glassan. They described 'a very strange object which was moving slowly on the flat, calm surface about eighty yards away'. It was reported as having 'a serpentlike head' and reckoned it to be over 6 m long. In the Irish Times of 7 October 2004, some of the mystery was laid to rest. The newspaper reported that a Swedish monster-hunter, Mr Jan Sundberg, who had carried out a search of the lake in 2001, believed that the monster is a vampire lamprey eel. He based this view on surveys his team had conducted on the lake and on interviews with locals, who had reported such phenomena as 'dead sheep and cattle along the shores emptied of all blood' and 'some kind of creature . . . more or less entangled around a cow'. Lamprey eels have no bone structures and breathe through gills.

There is a turn to the right as you exit Glassan that will lead you close to the shore of Killinure Lough and into Ballaghkeeran, where the Breensford river joins Lough Ree. The village's name is associated with St Ciaran, whom we mentioned earlier. From Ballaghkeeran, the main road branches off to the left at the Dog and Duck pub. To avoid using this busy road, take the road that climbs away from the pub, and turn right just before you are reunited with the N55. This route will lead you towards the last public shore-access point, on the lake at Coosan point. In making your way back to Athlone from Coosan point, you may wish to have a quick look at the aforementioned Lough Ree Yacht Club, which is located to the left down a narrow road leading to where the lough squeezes into the River Shannon.

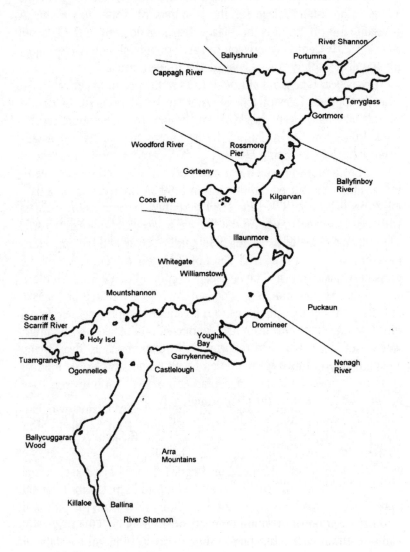

Lough Derg

(136 kilometres)

LOCATION
Stretches from Portumna in County Galway to Killaloe in County
Clare, with County Tipperary dominating its eastern shoreline and
Counties Galway and Clare sharing its western shoreline

LENGTH
39 km/24 miles

WIDTH
13 km at its widest point, but averages at 3 to 4 km/2 to 2.5 miles

AREA
12,800 hectares/12.8 km²

PUBLIC ACCESS
Western shore: Portumna harbour, Rossmore bay, Williamstown har-
bour, Church bay/Whitegate, Mountshannon, Knockaphort (Holy
Island), Scarriff, Tinarana bay, Twomilegate, Killaloe. Eastern shore:
Terryglass, Kilgarvan quay, Mota quay, Kilbarron quay, Luska pier,
Dromineer, Garrykennedy, Ballina.

MAP
Ordnance Survey of Ireland Discovery Series Map 53, covering Clare,
Galway, Offaly and Tipperary (ISBN 978-1-901496-55-0), Map 58,
covering Clare, Limerick and Tipperary (ISBN 978-1-903974-34-6),
and Map 59, covering Clare, Offaly and Tipperary (ISBN 978-1-
903974-35-3)

Lough Derg, described as Ireland's pleasure lake, is an excellent example of how to make the best use of a natural resource. It is the largest of the three great lakes along the River Shannon and by far the most interesting. From a cycling perspective, it is a delight. It has sufficient hills around its perimeter to afford excellent views, but none of the hills are too taxing to climb. Moreover, the lake's irregular shoreline provides plenty of variety, from forest parks to natural harbours and man-made quays. There are many historic sites to visit, and you will never be too far away from a village, where you can break your journey with suitable refreshment. The towns and villages dotted along the shore of Lough Derg take great pride in their appearance and have been the recipients of many national tidy-town awards over the years.

Origins of name

The lake derives its name from a most unusual tale. It brings us back to the time in Ireland when there were minstrels and poets wandering around entertaining the courts of the various chieftains and kings through whose territory they passed. The more successful ones had great power and influence – much like our modern-day pop stars. One such influential poet was an Ulsterman called Aithirne. He had the habit of making unusual demands of the kings and chiefs whom he entertained. On one occasion, he visited the court of Eochaidh Mac Luchta, king of South Connacht and Thomond, who resided near Mountshannon, on the western shore of Lough Derg, and who had only one eye. Aithirne requested that the king surrender his remaining eye to him. The king immediately complied, and went to the lakeshore to wipe the blood from his face. One of his entourage remarked that the water of the lake had turned red with the king's blood, whereupon the king declared that from that moment on the lake would be known as 'Loch Dearg Derc', or 'the Lake of the Red Eye'.

Description

Lough Derg is shaped like a reclining chair, with its broadest section stretching from Scarriff in the west to Dromineer in the east. Most of the rest of the lake is fairly narrow, with its average width being less than 4 km. The lake bed is predominantly limestone, except at its southern end, where the lake is sandwiched between Slieve Bernagh and the Arra Mountains, in an area of red sandstone and silurian rock. Originally it was dammed by a rocky ridge at Killaloe, but in the last Ice Age a glacier cut through this ridge, creating a narrow gorge, to allow the River Shannon to flow towards Limerick and the sea. The northern half of the lake is generally shallow and surrounded by low-lying countryside to the east and, in the distance, the Slieve Aughty Mountains to the west. In contrast, the southern half is deeper (up to 30 m) and more dramatic, with mountains reaching up from either shore. Between 1925 and 1929, a massive hydroelectricity power station was built at Ardnacrusha, south of Lough Derg, with the lake becoming the station's reservoir.

Lough Derg Cycleway

The Lough Derg Cycleway, an established 132 km cycling circuit around Lough Derg, is well signposted in both directions throughout. The route outlined in this book generally follows this circuit, with a few variations. A starting point is easy to choose. Those coming from the north will select Portumna, while those from the south will opt for Killaloe. For the purposes of this book, I have chosen Portumna. There are excellent parking facilities on the lakeshore at Castle harbour.

Portumna Castle

Either before you set off from Portumna or after your return, you should take the opportunity of visiting Portumna Castle. While described as a castle, this has more the appearance of a large mansion and is typical of a style of fortified house that developed around the end of the sixteenth century, replacing the more robust-looking castles. Portumna Castle was built around 1618 by Richard de Burgo within a demesne of 1,400 acres and was the main seat of the Clanricarde dynasty until it was accidentally burned in 1826. He was a descendant of William Fitzadelm de Burgo, who arrived in Ireland in 1185 with King John, married a daughter of Donal Mór O'Brien, King of Thomond, and was granted large tracts of land in Munster. King John appointed de Burgo's son, Richard, Lord of Connacht and granted him almost all the land of that province. The de Burgos acquired the earldom of Ulster through marriage, and it was said at the time that the family owned more land in Ireland than the king of England.

As the years passed, two distinct branches of the family developed in Connacht. One settled in the northern portion of the province and became ancestors to the Burkes of Mayo. The other took charge of the lands around Galway and established the Clanricarde branch. Initially, both branches of the family appeared to ignore allegiance to the English Crown, and their descendants ruled like Irish chieftains until the time of Henry VIII. However, in 1543, Ulick, who was then head of the Clanricarde branch, became the First Earl of Clanricarde, having submitted to the King. It was the Fourth Earl, Richard de Burgo, who built Portumna Castle, but it was his descendant Ulick John, the Fourteenth Earl, who had the more enduring impact on the castle and the family estate. Ulick John was born in 1802 and succeeded to the title when he was just six years old. In 1825, he married the daughter of George Canning, then British foreign secretary and later prime minister, and was quickly elevated to the title of Marquis of Clanricarde. In 1826, the year Portumna Castle burned down, he became Baron of Somerhill. He became a cabinet minister and was a member of Russell's administration during the Famine. While he did not resort to large-scale clearances of tenants during the Famine, he did little to alleviate the destitution of his tenants. For a long time after the fire in 1826, Portumna Castle remained in ruins. A new residence was under construction at the time of the Fourteenth Earl's death in 1874, but it was never fully completed and, like many houses of the aristocracy at that time, was burned down again in 1922.

In its present form, the fortified house has castellated tendencies, with square corner towers and decorative battlements. There is a large central block, with three storeys over a basement and a wide flight of stone steps leading up to the front door. As part of its fortifications, there is a large machiolation at parapet level over the doorway through which defenders could drop things on intruders. The circular porch was not original but was added in the eighteenth century. Restoration work on the building and gardens is ongoing. The ruins of a fifteenth-century Dominican priory stand nearby.

The western shoreline

The initial part of the journey from Portumna via the R352 takes you away from the lakeshore and is, frankly, uninspiring. Accordingly, I suggest that you start off by going through Portumna Forest Park, which you can access from Castle harbour. A new family cycling trail has recently been constructed through this forest. There is an exit from the forest on to the R352 about 1.5 km from Portumna. You will be out of sight of the lake for some time. In order to relieve the tedium and discomfort of travelling on a main road, take the first turn left after the forest park. This will lead you through a series of country lanes that run close to the lakeshore; you will eventually return to the R352 just short of the junction at Ballyshrule. Turn left at this junction to start the journey south. At the next junction, known as Power's Crossroads, you have the opportunity once again to leave the main road and embark on another loop, which brings you closer to the lake. For the more adventurous, it is possible to reach the lakeshore and track it all the way to Rossmore pier, but this will involve some off-road cycling and the use of a track that can at times be marshy and wet.

As you make your way back towards the R352, turn left after Rossmore Bridge in order to prolong your journey along the pleasantly quiet country lanes as far as Gorteen, where you have no option but to rejoin the main road. Just beyond the well-signposted boundary between Counties Galway and Clare, you can once more avail of the quieter country roads, by turning left almost immediately after the boundary line. Be careful not to take the left turn at the sharp bend just before the boundary, as this leads to a dead end. Quite early along this road, you cross over Derrainy Bridge. To the right of the bridge is Derrainy Kyle, an ancient burial site. The name 'Derrainy' is an anglicised version of the Irish 'doire aon oiche', or 'the oak wood that sprang up in one night'. The wood is located on the other side of the R352 along the route of a waymarked trail known as the East Clare Way. Passing a small lake called Lough Alewnaghta on your left, you make your way towards Williamstown harbour and skirt the lakeshore for a short while before returning inland to Whitegate.

Curiously, these loops off the R352 along the upper part of the western shore are not part of the official Lough Derg Cycleway. They afford the opportunity of using much quieter roads and also lead much closer to the lakeshore, in an area where views of the lake would otherwise be intermittent. In addition, once you pass Whitegate, you will have very little further opportunity to escape from the main road along the rest of the western shoreline. The loop around Whitegate to

Williamstown harbour is particularly attractive and yields rewarding views of the northern half of the lake just before it broadens out.

Holy Island (Inis Cealtra)

The pretty village of Mountshannon, located right on the lakeshore, is the next stop on the circuit. This village is very popular with boat-owners; those who do not own a boat can hire one at the pier in order to visit Holy Island (or Inis Cealtra), one of the most celebrated monastic sites in Ireland. On this island can be found a round tower, the ruins of six churches, an anchorite's cell or confessional, bullaun stones – which have bowl-shaped hollows etched on to them and are associated with both pagan and Christian rites – and high crosses, all packed into little more than nine hectares. The island's name is derived from the old Irish word for church, celtair, giving the name 'Church Island' if directly translated. Its popular modern name is more descriptive. For those interested in learning more about the lake, local historian Gerard Madden has written a comprehensive history entitled Holy Island: Jewel of the Lough. In the book, he describes how the island was regarded as sacred even in pre-Christian times. The first Christian settlement was established by St Colum in the sixth century, but the island is now mainly associated with St Caimin, who flourished in the seventh century and established the island as a centre of learning and prayer. He was abbot there for fourteen years. A later abbot on Holy Island was Marcan, the brother of the great Irish chieftain Brian Boru.

Edwin Wyndham-Quin (1812–71), Viscount Adare, later Third Earl of Dunraven, is said to have taken an ancient cross from the island for his museum at Adare. He erroneously believed that its inscription, 'Oroit do Chuinn' referred to a Quin ancestor. One of the great antiquarians of his time, George Petrie, who had alerted him to the existence of the cross, later advised him that the inscription should have read 'Oroit do Chunn', meaning 'a prayer for the soul of Choon'.

Holy Island, in particular its 22 m-high round tower, is a focal point in terms of the views across the lake, both from the western and eastern shores. However, as Madden suggests in his book, it is more rewarding to visit the island than to read about it.

Scarriff and Tuamgraney

From Mountshannon, the road skirts along Scarriff bay, from which there are fine views towards Slieve Bearnagh. Scarriff is a somewhat boring-looking village whose main claim to fame is that it was the location of an ironworks established in the seventeenth century by Ireland's then richest man, Richard Boyle, Earl of Cork. A new quay and marina facility has been established here by Waterways Ireland, so this might liven up the village. It is difficult to say where Scarriff ends and the next village of Tuamgraney starts, but the latter is a far more interesting proposition. A visit to the East Clare Heritage Centre, located in what is claimed to be Ireland's oldest church in continuous use, is a worthwhile diversion. There you can follow in the footsteps of Brian Boru, who would have crossed the threshold of the church, and learn about local history, myth and legend. The village name is an anglicised version of the Irish for 'the tomb of Gráinne', who is one of the most illustrious figures in Celtic mythology. The story of the elopement of Diarmuid and Gráinne is that of a greying husband who loses his wife to a younger and more attractive rival. It is one of the best-known and enduring Irish legends: it has sometimes been compared with that of Adonis and Aphrodite and may have been the inspiration for the romance of Tristan and Iseult. It is said that after the death of Diarmuid, Gráinne threw herself into Lough Graney, a lake a short distance west of Tuamgraney, and that her body floated down by what is now known as Scarriff river into Lough Derg. Locals claim that this is the first recorded suicide in Irish history.

The ruins of a fortified tower house stand beside the heritage centre. This is known as O'Grady's Castle and was the seat of the O'Grady clan, who were once a powerful family in the area. The castle is in private hands and is not open to the public.

After Tuamgraney, the R352 heads off to Ennis and you branch off on to the R463, which will bring you all the way to Killaloe. Be prepared for a steady climb away from Tuamgraney, but the consolation is improving views of the lake from your relatively elevated position. Just before the village of Ogonnelloe, there is an observation point from where the view across Scarriff bay is dominated by Holy Island. At Ogonnelloe, the main road turns sharply right and remains in close proximity to the lakeshore for most of the remaining ride to Killaloe. As you approach the most southerly part of the lake, there is an extensive parking and amenity area at Twomilegate, opposite Ballycuggaran Wood, where the University of Limerick has an activity centre.

Béal Boru ringfort

Before you arrive at Killaloe, you have the opportunity to visit the ringfort of Béal Boru, which is a couple of hundred metres off the road by the lakeshore. An information panel at the entrance provides some of its history of the ringfort. It was an important fortification of the O'Brien clan and is located at the point where Lough Derg officially finishes and the River Shannon once again emerges. It is now covered with mature trees but is still an impressive sight. The earthwork has a very large bank surrounded by a well-defined fosse or ditch. Its overall diameter is about 75 m. While it is now largely covered with grass, the bank would originally have been faced with stone.

Ringforts are circular spaces enclosed by a boundary of either earth or stone. The stone version is called a 'caher' or a 'cashel'. They were the dwelling places of farmers in the early Christian period (around 500 to 1000 AD) and are the most common of Ireland's surviving earthworks. It is estimated that as many as 40,000 ringforts still survive today. The Irish term for 'ringfort' is rath: a glance at a map reveals how many townlands in Ireland feature the word as a prefix to their names. A ringfort usually took the form of a circular habitation site with a deep outer fosse or ditch, and high embankments. Some ringforts had up to three fosses, and it is thought that the greater number was an indication of their relative important status.

Although Béal Boru is described in most of the tourist literature as a ringfort, David Sweetman, in his book The Medieval Castles of Ireland, is of the opinion that the structure is almost certainly a ringwork castle, which is built on a pre-existing enclosure, rather than a ringfort. He bases this view on the fact that the banks of a ringwork are more pronounced, and the fosse is wider, than one would expect to see on a ringfort. In addition, he states that 'the entrance to a ringwork is also distinguishable from a ringfort in that it will often have a pronounced ramp, and each side of the gap in the rampart will be faced with stone'.

Before the dredging of the River Shannon, which commenced in the 1840s, the river was fordable at this point. It was here that General Patrick Sarsfield, who led the Irish cavalry at Limerick during the

Jacobite War in August 1690, crossed the river on his way to a daring and ultimately successful raid on a military siege train belonging to King William's army that was on its way to Limerick.

Killaloe

Killaloe is an attractive town with a great deal of history behind it, given its pivotal position on the River Shannon. Its name is derived from the Irish 'Cill da Lua', which translates as 'Lua's Church', and refers to St Lua, who lived in a monastic settlement on a small island in the River Shannon about 1 km downstream from the town. The town was the birthplace of one of Ireland's greatest high kings, Brian Boru, who reigned from 1002 to 1014, when he was killed at the Battle of Clontarf in Dublin in the course of a strategic victory against the Vikings. His palace was located at Kincora, where Killaloe now stands: during Brian Boru's reign it was the seat of government and the capital of Ireland. It was destroyed in the twelfth century by Turlough O'Connor, King of Connacht, who razed it and threw its stones and timber into the River Shannon, leaving no trace. At the town's heritage centre, located within the Tourist Information Office on the west side of the narrow bridge that links Killaloe and Ballina, there is an exhibition covering the life of Brian Boru and the history of the surrounding area.

On the approach to Killaloe, you pass alongside part of the Killaloe to Limerick Canal. This was built in 1799 and allowed river traffic to pass through Killaloe for the first time. The River Shannon drops over 30 m in the 25 km between Killaloe and Limerick; prior to construction of the canal, a combination of shallow waters and rapids prevented navigation to Limerick. The canal allowed commercial traffic to move up and down the river, but railways and improvements in road networks eventually sounded its death knell as a commercial artery, and the last commercial load was transported on the river in 1959. Today it offers cruising boats a passage to Limerick and the Shannon Estuary.

St Flannan's Cathedral

Another interesting link with history in Killaloe is St Flannan's Cathedral, which was originally built by Donal Mór O'Brien, king of Munster, around 1180. Soon afterwards, the building was destroyed by

Cathal Carrach of Connaught, but it was rebuilt in the thirteenth century and the Romanesque doorway from the original was incorporated into the south wall of the new structure. Inside the entrance is a large stone on a pedestal that was discovered in 1916 in the wall surrounding the grounds of the cathedral. It is thought to have been a part of a cross, but in most literature about the cathedral it is described as an ogham stone because of carvings found on the stone which are in ogham, the earliest-known Irish script. These inscriptions refer to somebody called Thorgrim, who is thought to have been a monk of Viking origin who lived in these parts more than a thousand years ago. The small, roofed building in the cathedral grounds is known as St Flannan's Oratory. A more interesting oratory is to be found in the grounds of the Catholic church at the top of the town. This is known as St Lua's Oratory and is said to date back to the ninth or tenth centuries, but it did not always stand in its present location. When the Shannon Hydroelectric Scheme was being commissioned in 1930, it was realised that the island on which St Lua had lived, south of Killaloe, would have been submerged. Accordingly, the oratory was taken down and re-erected on its current site.

Crossing the stone bridge, with its thirteen arches, you leave County Clare and enter County Tipperary and the town of Ballina. When it was first erected, during the reign of Queen Elizabeth I, this bridge had nineteen arches, but six arches have been lost during repairs and rebuilding work carried out over the years. The bridge is very narrow, with no provision for pedestrians, except for several embrasures on the Clare side, which pedestrians can slip into to avoid passing traffic. It was announced in February 2006 that a site had been chosen for a new bridge to link Killaloe and Ballina in order to relieve the congestion at the existing bridge. The new bridge will be located between Cloonfada and Roolagh, a little less than 2 km south of the existing bridge. At the eastern end of the existing bridge, there is a fine bar and restaurant called Molly's, which is worth a visit.

The eastern shoreline

The R494 road climbs away from Ballina, in the lee of the Arra Mountains. These mountains are famous for their slate quarries and for the Bronze Age burial chamber found on their highest peak, Tountinna. The chamber has a line of small stone slabs which are known as the 'Graves of the Leinstermen' and, according to local legend, are the burial place of a minor Leinster king and his small army, who were slain by Brian Boru's wife while on their way to a wedding in Killaloe. Approximately 9 km from Ballina there is a parking facility known as the 'Lookout', which provides great views across the lake north and south, as well as right across it at its widest point. Immediately underneath the Lookout is the ruin of Castletown Church and burial centre. A short ride north of the Lookout, you will be able to take a left turn to leave the R494 and rapidly descend to the lakeshore, past Castletown Church and on to Castlelough Lakeside Park. It is a pleasure once again to avail of quiet country lanes, even though some have the signs of being well trafficked by livestock and are quite heavily soiled. Castlelough Woods, to the left, offer a 3.5 km forest walking trail on the border of the lake, should you be feeling especially energetic.

The Lough Derg Way

You are now in an area where you can easily go astray. On the first occasion I did the Lough Derg circuit, I was fortunate in that I bumped into a number of locals, who advised me to follow the way-marked trail signposted as the Lough Derg Way rather than the cycle-way. This meant that in order to get to Garrykennedy, I did not have to make my way back to the R494, which would have been tiresome and also would have flown in the face of my guiding principle of keeping as close as possible to the lakeshore. Indeed, on that occasion my good fortune was such that a local farmer, whom I encountered driving a tractor, told me to follow behind him. He made his way through several gateways and across the fringes of a number of fields, all on the route of the Lough Derg Way, before linking up with the cycleway less than 1 km west of Garrykennedy. I would strongly recommend pursuing this route, as it knocks quite a few kilometres off the journey, and the views are much nicer. In addition, you save yourself repeating a steep climb, as Garrykennedy lies at the edge of a shore that is bound to the south by rising ground.

Garrykennedy

The village of Garrykennedy, looking out on to Youghal bay, is very inviting and is a busy boating location during the cruising season. The small stone harbour was once used for the transport of slate quarried from the nearby Arra Mountains. The ruins of a fifteenth-century tower-house stand on the harbour wall. This was built by the local landowners, the O'Kennedy family. Also known as Castlegare and Slanger Castle, it is only a shadow of its former self, with much of its stone having been used to construct the harbour.

There is a steep climb out of Garrykennedy, and at the first opportunity you should turn left once again, following the route of the Lough Derg Cycleway. This road leads towards the village of Newtown, where you once again link up with the R494, albeit for a very brief period. On your way to Newtown you can, if you wish, divert off to Youghal village and harbour. Do not be tempted to follow the Lough Derg Way, which splits off from the cycleway just after the turn off to Youghal, as this will lead you to an area which is best left to walkers – unless you wish to carry your bike along some challenging stretches.

Dromineer

After Newtown, head for Ballycommon, but there is no need to go all the way to the village, which would mean linking up with the main road again. Instead, turn left at the first opportunity and then right, allowing you to bypass Ballycommon. Eventually, you will find your way on to the R495 for a brief run before branching off on to a minor road, which provides an alternative and quieter route to Dromineer.

Dromineer was an important port when commercial traffic navigated the Shannon. It is the home of the Lough Derg Yacht Club, founded in 1836: this club is not quite as old as its Lough Ree counterpart but is legendary all the same among Ireland's sailing community. The village is also a bustling centre for water sports. For those with a focus on the shore, it has, like Garykennedy, the ruins of a fifteenth-century towerhouse, and also has the ruins of a tenth-century parish church.

Leaving Dromineer by the R495, the next target is the village of Puckaun. Follow the cycleway signposts, crossing over the Nenagh river by the Annaghbeg Bridge. In this area, you will come across signs for Slí Eala, featuring the emblem of a swan. This is a marked walking route along the banks of the Nenagh river, which flows into Dromineer bay. Puckaun is unremarkable except for a pretty and tasteful development of Irish holiday cottages in the middle of the village. From Puckaun, continue north on the R493 until you come to a signpost for Luska bay, where you turn left to embark on a loop that will bring you right to the lakeshore. Here there are excellent views of Illaunmore, the largest island on Lough Derg, and the smaller Cameron Island. Halfway through this loop, you cross over a small river called the Ballycolliton, which rises from two springs in an area called Springmount a couple of kilometres to the east. 'As clear as the Ballycolliton' is an old local saying in this area. The loop continues on to Kilbarron quay and then climbs back to the R493 at Coolbaun.

Terryglass

As you journey northwards, there are several shore-access points off the R493, including Mota quay, quickly followed by Kilgarvan quay. Before you reach the village of Ballinderry, turn left to avail once again of the quieter minor roads and to get closer to the lakeshore. At the southern entrance to the village of Terryglass, there is a gateway which leads to a large, unfinished thirteenth-century Norman castle called Oldcourt Castle. The roadside entrance is closed, but you can get a view of the castle from the quayside, which is signposted at the centre of the village. Also in the village are the remains of a sixth-century monastery founded by St Colman. The Book of Leinster, which is on display at Trinity College in Dublin, was written at this monastery between the twelfth and thirteenth centuries. The abbey church was destroyed by fire in 1164.

As was the case for the initial part of the journey away from Portumna, the remainder of the circuit is uninspiring. Once you turn on to the N65 at Carrigahorig, you have to contend with fast-moving traffic on a relatively narrow main road with no hard shoulder. Before you reach the town, you cross the large bridge over the Shannon on the outskirts of the town. The bridge was constructed with a lifting section, and a cottage was built on the bridge for the bridge operator, who could respond to signals from the larger boats. Waterways Ireland propose to install floating moorings downstream of the bridge, along with a 20 m floating breakwater upstream and a 33 m walkway from the bridge. It is arguable that these developments might take away from the existing pleasant and peaceful views of the river from the bridge.

County Leitrim

County Leitrim

Lough Melvin
Lough Macnean Upper & Lower

The county of Leitrim originally formed part of the old Irish kingdom of Breffni, which was ruled by a powerful Irish clan, the O'Rourkes. A Norman invasion of the kingdom in the thirteenth century was resisted in the northern region, which remained in the hands of the O'Rourkes until the sixteenth century. At that time, large parts of the county were planted with English settlers, but the plantation was largely unsuccessful, probably because of the poor quality of the land. In 1583, the Lord Deputy, Sir John Perrott, marked out the boundaries of what now constitutes the county, with the principal town being Carrick-on-Shannon.

The county name, and that of one of its villages, is an anglicisation of the Irish 'Liath Druim', which translates as 'the Grey Ridge'. It is a common place name in Ireland, with over forty other Leitrims to be found as townlands, villages or streets across the country. The county was very badly affected by the Great Famine of the late 1840s. Evidence of the impact of depopulation is to be found throughout the county, in the form of the many ruined cottages and houses that you will encounter on your travels around its lakes.

Leitrim's landscape is pockmarked with many lakes, the majority of which are small. However, it does boast one large lake at its centre, Lough Allen (which has already been covered with the other great lakes of the River Shannon, see page 99). The lakes included in this section do not fall entirely within the county's boundaries. They are shared with County Fermanagh and form part of the border between Northern Ireland and the Republic. Because of their location in the border region, circuits around the lakes would have been difficult up until recently. The minor roads around their shores were classified as unapproved roads and were blown up, creating unpassable blockages, to prevent unsupervised border crossings. Fortunately, all obstacles have now been cleared, and the roads have been resurfaced, providing comfortable and accessible routes that are largely undiscovered.

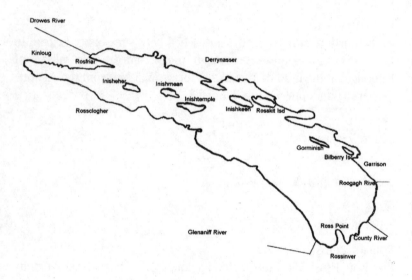

Drowes River

Kinloug

Rosfriar

Derrynasser

Inisheher

Inishmean

Inishtemple

Rossclogher

Inishkeen   Rosskit Isd

Gorminish

Bilberry Is

Garrison

Roogagh River

Ross Point

Glenaniff River

County River

Rossinver

Lough Melvin

(38 kilometres)

LOCATION
In the north-west of Ireland, sandwiched between County Leitrim to
the south and County Fermanagh to the north. The village of
Kinlough, at the head of the lake, is located 6.5 km from the Atlantic
Ocean at Bundoran, County Donegal.

LENGTH
12 km/7 miles

WIDTH
2.4 km/1.5 miles at its widest point

AREA
2,125 hectares/23 km^2

PUBLIC ACCESS
Garrison, Rossinver, Breffni pier, Kinlough pier, Derrynasser Amenity
Area, Garvros

MAP
Ordnance Survey of Ireland Discovery Series Map 16, covering
Donegal, Fermanagh, Leitrim and Sligo (ISBN 978-1-901496-08-6)

If one were to rank the lake circuits covered in this book, the Lough Melvin circuit would certainly rank in my top three. Except for anglers and locals, this lake is little known, yet to those in the know it is special. I am not alone in my admiration for its attractions. The following is a description dating back to the late nineteenth century by Canadian-born Sidney Berdroe Wilkinson, who arrived in Ireland in 1867 as a member of a British Government Survey and was stationed for some time at Castlebar but visited the shores of Lough Melvin regularly:

> The whole surface of the lough sparkling and scintillating with blue, violet, green and brown and amber, where the shallows showed the lovely delicate tints of the young larch and the catkins on the hazel bushes, young shoots of ferns and wild flowers on the islands.

Little has changed. There is an unspoilt quality about its shoreline, with the Dartry mountain range providing a dramatic backdrop along its southern shore. The roads that are used for the circuit were closed to traffic until recently, and this has probably contributed to the lack of awareness of its attractions as a lake circuit. Today, they afford comfortable and sedate passage, with sufficient elevation to allow for appealing views of the lake for virtually the entire circuit.

Legendary origins

The origins of the name 'Lough Melvin' are unclear. An older name for the lake is 'Lough Melve', although who or what 'Melve' was is unknown. It has also been suggested that the lake's name is an anglicisation of the Irish maolbhean, or 'the bald woman'. This reference is associated with one of the mythical tales surrounding the origins of the lake. Local folklore relates the story of an old woman from Belleek who came to Kinlough, which lies at the lake's northern head, to draw water from a well. When she was leaving, she forgot to replace the lid on the well, with the result that the water followed her on her return journey, creating the lake. It is not explained why the lake is somewhat distant from Belleek and has a more southerly profile than would be justified by the route of her return journey.

Unique fish species

References in this book to fishing are few and far between, mainly due to my lack of knowledge on the subject. However, it would be doing

Lough Melvin a great disservice not to mention its unique status in the fishing world and its boast of being the best game-fishing lake in Ireland. It is renowned as one of the few examples of a post-glacial salmonid lake boasting genetically distinct species of trout not to be found elsewhere in the world. These are the gillaroo (Salmo stomachius), a trout with vivid red spots whose name is derived from the Irish gille ruaidh, meaning 'red fellow', and the Sonaghan (Salmo nigripinnis), noted for its black fins. The lake is also home to another trout species called the ferox (Salmo ferox), whose local name is 'bracklough' and which is probably related to the brown trout of Lough Erne. For information on this, I am indebted to the research work of Professor Andrew Ferguson of Queen's University Belfast, who has studied this lake over many years. In November 1985, he delivered a lecture at the Royal Dublin Society on the results of a study he had conducted to determine the exact status and inter-relationships of these unique species and their relationship to brown-trout populations elsewhere. He reported that these trout have their origins in former migratory sea-trout populations and may have been present in this lake for up to ten thousand years, when the ice retreated at the end of the last Ice Age. They are a purely natural species and are not the result of stocking, and for this reason are of great interest to anglers, whose fishing endeavours in most waters in Ireland focus on a single universal species, the brown trout.

It is surprising, given the size of Lough Melvin, that there has been little interbreeding between these three species, but this is due to the fact that each species has a distinct spawning ground. The gillaroo spawn in the lake margins around Lareen bay, near the outflowing Drowes river, while the Sonaghan spawn in in-flowing rivers such as the Ballagh and Tullymore. The ferox is the least common of the trout to be found in the lake, and its spawning grounds lie in the deep trench starting near Rossinver bay, where the Glenaniff river flows into the lake. The lake also yields salmon in the early spring, and during the summer months grilse (called 'grawl' locally) are a regular catch. The lake is a veritable angler's paradise. It is interesting to note that in Sidney Wilkinson's time, fishing was prohibited on a Sunday; there are of course no such restrictions today.

Description

Structurally, there is a marked difference between Lough Melvin's northern and southern shorelines. The southern shore is dominated by the presence of the Dartry mountain range, which runs along almost its entire length. Many streams and watercourses run down the mountain slopes and into the lake. The shoreline runs fairly straight, with few coves or outcrops of land. The northern side is a complete contrast, with many bays, coves and promontories. Most of the lake is between 10 m and 20 m deep, with shallower areas near its six main islands and at the western end, as it narrows towards Kinlough. There is a deep trench, with a maximum depth of 45 m, running east to west from Rossinver bay in the south-east corner to Lareen bay in the north-west. This trench is effectively a continuation of the steep gradient of the Dartry mountain slopes.

Dobharchú

One of the more interesting tales about Irish lake monsters relates to Glenade Lough, a 200 acre lake which lies close to Lough Melvin in County Leitrim, approximately 8 km north-west of Manorhamilton. The lake's monster is known as the 'Dobharchú'. The fearsome beast was said to have been half-wolf-dog and half-fish, and like an enormous otter. There is a lengthy poem that relates the tale of how the 'water fiend' emerged from the lake in 1722 to devour a lady called Grace Connolly while she was bathing. The full text of this poem is on display at the Dartry Delights Coffee Shop in Kinlough, on the northern shore of Lough Melvin. Local historian Joe McGowan, author of Echoes of a Savage Land, relates one version of the tale, about the demise of Grace Connolly. When Grace failed to return from the lake, her husband, Traolach Mac Lochlainn, went in search of her. He found her dead on the shore of the lake, with the beast lying across her. Mac Lochlainn shot and killed the beast, but as he did so a second animal emerged from the water and chased him. There was a struggle near Castlegarden Hill, and Mac Lochlainn managed to subdue and kill the second beast, and buried it where it fell. Like most legends, there are various versions of this tale: one of these suggests that Grace Connolly was in fact murdered and that the story of the Dobharchú was a cover-up. To add authenticity to the monster story, her tombstone, at nearby Conwal cemetery, clearly depicts an otter-shaped beast being killed.

Garrison

The circuit can be started at either of the two villages located at the eastern and western extremes of the lough: Garrison and Kinlough, respectively. For the purposes of this book, I have chosen to start at Garrison, whose name has been attributed to the establishment of a military barracks here by King William III after the Battle of Aughrim. Mr Wilkinson described it in the nineteenth century as a 'delightful spot', and it retains that image. It is sited at the mouth of the Roogagh river, and for the more adventurous there is a short trek to a waterfall signposted off the B52 as you enter the village. The Lough Melvin Holiday Centre, located within the village, is a residential hostel with a small restaurant and camping-and-caravan facilities to the rear. The centre can also provide bicycles for hire. There is a spacious car park and amenity area located on the lakeshore on the outskirts of the village, where you can even avail of a shower room.

Leaving Garrison in a southerly direction for a clockwise circuit, you pass over the unusually named County river, the first of many rivers you will encounter on your journey. There is a shore-access point just pass Rossinver Fishery; as the road starts to turn away from the lake, turn right on to the R281, the Rossinver road. After crossing the Glenaniff river, there is a prolonged but steady climb, which keeps parallel to the shoreline. As the road flattens out, there are beautiful views across the broader western end of the lake, across to the heavily wooded Gorminish Island. The translation of this name, from the Irish, is 'Blue Island' – one of the colours mentioned in Mr Wilkinson's description above.

The southern shoreline

The road along the southern shore is well surfaced, and the sound of running water is never far away, with lots of streams running off the mountain into the lake. There are uninterrupted views across the lake, and one wonders how long it will be before houses start to appear on the elevated plots between the road and the lakeshore. At present, you will encounter several dilapidated structures along the roadside: I am sure that it will not be long before these are refurbished.

A little under halfway along the southern shore is Breffni pier, from where you can look across to nearby Inishtemple – or Church Island, as it is more commonly known locally, because of the ruins of a fifteenth-century church which are to be found on the island. After Breffni pier, the road moves inland, rising steadily and edging closer to the mountains. Your attention will be absorbed by the dominant Arroo Mountain, whose summit is often shrouded in mist, and which features several strangely shaped rocky outcrops. The stone structure that you can just about make out when the mist clears from the summit of Arroo Mountain is called the Sapper's Mark. You will get a better view of this later on in the circuit, from the opposite shoreline.

Rossclogher Castle

As you descend towards Kinlough, a signpost indicating the presence of Rossclogher Castle and its links to the Spanish Armada will bring you to a halt. The castle was a stronghold of the MacClancy clan who were sub-chieftains of the principal clan in this region, the O'Rourkes of Breffni. Rossclogher Castle cannot be seen from the road, as it is actually sited on a small island about 80 m from the lakeshore. In order to view it, access to the shore can be obtained via the rough lane to the right of the road; this lane leads through a number of gates to an elevated field. There you will find not only a good position from which to view the castle ruins but also the remains of a small church known as 'Doire Meala', which was possibly built by St Tiarnach in memory of his mother St Meala. There is also a small ringfort on higher ground above the church ruins, indicating the importance of this site even before the castle was constructed in the early fifteenth century. This elevated position provides the best views of Lough Melvin and its islands, a number of which have been purchased in recent years by Charles Ferguson, a local man who emigrated many years ago and, having made his fortune overseas, has returned to the area and invested in the local heritage.

Rossclogher Castle, whose name is derived from the Irish 'Ros Clochair', meaning 'the Point of the Stony Place', flourished from around 1400 to 1641. It is hard to gauge how large the castle was, as there are few remains other than an ivy-clad tower.

The Spanish Armada

The castle's connection with the Spanish Armada is due to a Spanish captain called Francisco De Cúellar, who survived being shipwrecked in Donegal bay and landed at Streedagh beach in County Sligo. In 1588, a fleet of 130 Spanish ships of various sizes and types attempted to invade England, but failed due to bad weather and English counter-attack. The fleet was forced to flee north, and it is estimated that twenty-five ships foundered off the coast of Ireland between north Antrim/Donegal on the northern coastline and Blasket Sound off County Kerry in the south. Three of the ships, the La Juliana, the La Livia and the Santa Maria de la Visōn were wrecked at Streedagh. De Cúellar had been on board the La Livia, having earlier been relieved of command of his own ship, the San Pedro, for not adhering to the fleet formation. Despite not being able to swim, he and eight fellow survivors scrambled to the shore. They made their way overland from Sligo to Antrim, in order to return to Spain. It was during this journey that De Cúellar found his way to Rossclogher, where he was welcomed by the Mac Clancys. This was unusual, as the Irish seldom offered help or refuge to the suvivors of the destruction of the Spanish Armada. However, an English force of 1,700 men under Lord Deputy Fitzwilliam had been detailed to search for survivors, and they arrived at Lough Melvin not long after De Cúellar had reached the castle. Strangely, Mac Clancy and his men took to the hills before the English arrived, and left it to the Spaniards to defend the castle. For three weeks, the English laid siege to the castle, without success; then the weather played its part, with snow forcing the English to abandon their positions. Mac Clancy was very grateful to De Cúellar, and urged him to stay on, but the Spaniard was eager to return to his homeland, and on 4 January 1589 he slipped away quietly, eventually making his way back to Spain via Antwerp. He described his experiences in Ireland in a letter, dated 4 October 1589, to King Philip II of Spain; the letter is one of the few surviving accounts of the ill-fated Spanish Armada.

Kinlough

As you approach Kinlough, you will notice the ruins of a large house to your left. These are the remains of Oakfield House, formerly owned by the Foley family. Unlike other large country houses which succumbed to fire around the time of the Irish civil war, this house survived, but the owners de-roofed it in the middle of the twentieth century to avoid paying rates – which at that time were very high, and a huge drain on the coffers of landowning families. Kinlough pier is located to the right as you enter the village from the Rossinver road. Kinlough is a friendly village, and you should not pass through without visiting the Dartry Delights Coffee Shop, which, as well as serving tasty meals, houses a small exhibition focusing on the local area and its native industries.

The Annals of the Four Masters

As you leave the village in the direction of Bundoran, watch out for the well-concealed turn to the right that leads down a quiet country lane running parallel to the lakeshore towards Lareen bay and Rosfriar point. The road turns sharply and links up with the Drowes river, a vigorous watercourse that drains Lough Melvin into Donegal bay. Mounted on the parapets of Mullinaleck Bridge, which spans the Drowes, is a bronze memorial to the authors of the Annals of the Four Masters – or, to give them their correct title, Annals Rioghachta Éireann/the Annals of the Kingdom of Ireland. The Annals, an important source of Irish history, were said to have been compiled during the period 1682–86 by a Franciscan Brother, Micheál Ó Cleirigh, assisted by three lay associates, Peregrine Ó Duigneáin of Leitrim, Peregrine Ó Cleirigh of Donegal and Fearfeasa Ó Mulcoonry of Roscommon. The memorial says that the Annals were compiled along the banks of the Drowes in a house of the Friars of Donegal. There is no indication as to where that house was, but the area name, Rosfriar, is an anglicised version of the Irish 'Ros an Bhrathair', as if to confirm the association. Unlike the Annals of Ulster, which begins with the mission of Palladius to Ireland in 431, the history outlined in the Annals of the Four Masters goes back much further in time, and in its earliest descriptions mixes myth and history. It finishes in 1616.

Not too far from Mullinaleck Bridge, you will need to turn right just after the entrance to the Drowes Fishery. There is very little traffic on these country lanes, allowing you to enjoy excellent views across to the Dartry Mountains in almost perfect tranquillity. A sharp turn to the right is aptly located in an area called Askill, an anglicisation of the Irish word ascaill, meaning 'corner', and reflects the shape of the townland as it angles its way towards Rusheen point. Another nearby promontory is called Burke's point: it was from here that the locals used to bring their dead for burial on Inishtemple, mentioned earlier.

Crossing the 'border'

At Derrynasser amenity area, you get a fine view of Inishkeen, which means 'beautiful island'. This is also known as Maguire's Island, after a Doctor Maguire who used to live on it. It is aptly named, and amidst the thick woodland you will be able to see the ruins of Doctor Maguire's house on the shore facing the mainland. Abhornaleha Bridge, just past Derrynasser, marks the border between Northern Ireland and the Republic. A memorial on the side of the road near the bridge is a reminder of the Troubles that plagued the province of Ireland in the last decades of the twentieth century. The memorial marks the spot where a local man, John Dolan, drowned in February 1994, in a waterlogged crater that was left after the British army had blown up the road to ensure that it could not be used for unsupervised travel between the two jurisdictions. Fortunately, the road is now open once again, but the fact that it was closed for almost three decades means that some people living close to the border would not know their neighbours, despite living very close to them. A more positive legacy of this episode is the fact that the surface of the road on the northern side of the border has been improved, and provides an extremely comfortable cycling surface, which makes the undulating route towards Garrison easier to bear. Just before you reach Garrison, there is another fine shore-access facility at Garvros, which looks on to Bilberry Island.

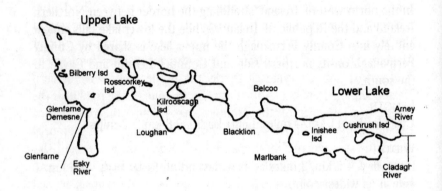

Lough Macnean Upper & Lower

LOCATION
In the north-west of Ireland, straddling the border between Northern Ireland and the Republic of Ireland. While the lower lake falls almost entirely into County Fermanagh the upper lake is shared by County Fermanagh on its northern side and Counties Leitrim and Cavan to the south.

LENGTH
Upper lough – 11 km/7 miles. Lower lough – 2.8 km/1.7 miles.

WIDTH
Upper lough – 5 km/3 miles at its widest point. Lower lough – 1.6 km/1 mile at its widest point.

AREA
1,013 hectares (for both loughs combined)

PUBLIC ACCESS
Upper lough – Lough Macnean Park, Glenfarne demesne, Meenagh. Lower lough – No official access point, but there is a car park, and walks, along the lakeshore at Belcoo.

MAP
Ordnance Survey of Ireland Discovery Series Map 26, covering Cavan, Fermanagh, Leitrim, Roscommon and Sligo (ISBN 978-1-901496-48-2), and the Official Route Map for the Belfast–Ballyshannon cycle route, produced by the National Cycle Network, available from the National Cycle Network Information Service (Tel. 0044 117 929 0888) or from Sustrans (Northern Ireland), McAvoy House, 17A Ormeau Avenue, Belfast BT2 8HD, Tel. 0044 28 9043 4569 (ISBN 978-1-901389-21-0)

Lough Macnean is little known outside the locality, except to intrepid anglers. In terms of promotion, it is often neglected, and I have even come across an activity guide for the region produced by a state agency where the lake does not even merit inclusion on the map! In my view, this is a serious omission, because a significant section of the upper lake has one of the most appealing off-road lakeshore trails to be found anywhere in Ireland.

Description

Structurally, the lake has two distinct elements, linked by the Belcoo river. The larger upper lough lies to the west and, despite its small area, is shared by three counties: Cavan, Leitrim and Fermanagh. The lower lough lies almost entirely in County Fermanagh, with a small section in County Cavan, near Blacklion. Both lakes are spring-fed and set in an attractive, rugged landscape. What the lower lough lacks in size, it makes up for in terms of its dramatic surroundings. The Cuilcagh Mountains, source of the River Shannon, dominate to the south, while Belmore Mountain rises to the north. Both elements of the lake are fairly deep, with the upper lough reaching depths of 21 m, while the lower lough has a maximum depth of 12 m. The lakes are drained to the east by the Arney river.

Border villages of Belcoo and Blacklion

As there is a natural division between the two parts of Lough Macnean, I have chosen to divide the cycling circuit into its upper and lower components. The starting point for each is either of the two villages that lie at opposite ends of the bridge over the Belcoo river. (The river marks the county boundary between Fermanagh and Cavan and the border between Northern Ireland and the Republic.) Belcoo, on the northern side, is of relatively recent origin, owing its existence to the Enniskillen–Sligo railway line, constructed in the 1870s. The line is now dismantled, and evidence of its course will be encountered at various locations around the shores of both the upper and lower loughs. The village name is derived from the Irish 'Béal Cú', which translates as 'Mouth of the Hound'. Despite the village's relatively brief existence, its nearby hinterland has a lengthy history, aspects of which will be revealed during the course of the circuit. The village on the southern side of the bridge, Blacklion, owes its name to a pub called the 'Old Black Lion Inn', which is now called the 'Diamond Bar'. This stood on what was an old coach road running through the area. It is said that there was another village further along this road named after a pub called the 'Red Lion' but that it became depopulated during Ireland's Great Famine of the 1840s.

Upper Lough Macnean

(34 kilometres)

This circuit has an inauspicious start, as the exit from Blacklion in a westerly direction has to be undertaken on the N16, until you reach the village of Glenfarne. Views of the lake are intermittent, although there is a car park and amenity area at Lough Macnean Park, just under 2 km from Blacklion. Along the road you will see signs indicating that you have entered 'Cathal Buí Country'. These refer to a celebrated poet called Cathal Buidhe Mac Giolla Ghunna (1670–1750), who was born at Tullyhaw near Blacklion. (The word buí is the Irish for 'yellow'.) There is a small monument to the poet near the ruins of Killinagh Church, which are surrounded by a twelfth-century graveyard. The ruins can be accessed across a farmer's field, approximately 100 m off the road to the right.

Once you pass through the small village of Glenfarne, bear right on to the R281, opposite the Ballroom of Romance, said to have inspired the novelist William Trevor to write the book of the same name. The building was erected by John McGivern in 1934 and, while showing signs of its age, still appears to be in regular use. The abutments of the bridge that once bore the Enniskillen–Sligo railway line still stand by the roadside, but the bridge has long since been removed. The small former station house can be seen to the left. It has been sympathetically restored and is now a private dwelling.

Glenfarne demesne

The highlight of this circuit is the trail through Glenfarne demesne, a Coillte facility that is accessed off the R281 just after the bridge over the Glenfarne river. Glenfarne means 'valley of the alders', and the demesne was once the home of the Tottenham family. The history associated with the Tottenhams, like that of many landlord families, is both good and bad. The first landlord, Nicholas Loftus Tottenham, fell out of favour locally when he sequestered land along the lakeshore, throwing many families off the land in the process. His descendants appear to have been more benign, building villages such as Kiltyclogher and providing food to the workhouse at Manorhamilton. The man who was eventually to lose the estate was Arthur Tottenham, who was Member of Parliament for Leitrim from 1880 to 1885. He got into heavy debt through his involvement with the construction of the Enniskillen–Sligo line and sold the estate to Sir Edward Harland of Harland & Wolff shipyards in Belfast. Glenfarne House met the same fate as many of the great houses of Ireland during the Civil War and was eventually deserted and fell into ruins. There is a story told locally that when the owners first deserted the house, the locals looted it, taking whatever furniture, fittings and artworks remained. The following Sunday, the local priest preached a sermon in which he vociferously condemned thievery and looting – without making any reference to recent local events. Such was the impact of his sermon that on the following Monday morning, there was a queue of carts bringing back the items that had been taken from the house.

Once you enter the demesne, there is a 2 km cycle on a metalled surface to an amenity and recreation site located on the shores of the lake. From there, walkers have the opportunity to avail of a superb trail along the lakeshore: it stretches for 5 km, with excellent views down the full length of the lake. Several islands lie close to the shore, and there are a number of resting areas where the lapping waters and bird-song are the only sounds you can hear. The first resting area you will encounter is called 'Ladies Rest': it is thought to have been a popular resting point for the Tottenham ladies as they walked around the demesne. Fortunately, it is now available to a much wider body of people. This is one of the most tranquil and attractive areas that I have visited along the shores of Irish lakes.

Lough Macnean Sculpture Trail

As you travel along the lakeshore, you will encounter several stunning sculptures. These form part of the Lough Macnean Sculpture Trail, a collection of eleven pieces located around the Lough Macnean valley area. They are intended to focus on the rebuilding of links between the communities around the shores, which, as happened in the vicinity of Lough Melvin, were disturbed by the Troubles. One of the more interesting sculptures you will come across in the Glenfarne demesne is not part of the sculpture trail. Located at the edges of the demesne is a piece entitled 'New Beginnings 2000', created by women from Belfast, Dublin and Leitrim, under the direction of Jackie McKenna and Brigid Corcoran. It consists of three standing stones with various bronze inserts depicting images from nature, such as birds, fish and trees, together with a stone seat where you can sit and contemplate.

The Black Pig's Dyke

There is no formal exit from the demesne, but the return to civilisation is marked by a reversion to a metalled surface as you pass the shore-access point at Meenagh. Continuing straight along this narrow lane will return you to the Glenfarne–Kiltyclogher road, the R281. Just before you reach this road, there is a signposted diversion that you can take to view the earthworks associated with the Black Pig's Dyke. This was a defensive structure erected between the ancient provinces of Connacht and Ulster around the first century BC. It consisted of a 3.7 m-deep ditch and a 6 m-high rampart supplemented by the natural boundaries provided by lakes, rivers and hills. Remnants of the ditch are to be found scattered throughout the border regions. It is interesting to view this ancient barrier in the context of the disputed territory of Ulster in modern history. Those of a unionist persuasion can call upon its existence as justifying the political separation of Ulster from the rest of Ireland. (A complicating factor is that the ancient ditch extended much further south than the current boundaries of Northern Ireland, however.) Traces of the ditch have been found on the shores of Lough Gowna near the townland of Dring and near Lough Ramor in Cavan.

As you emerge on to the Kiltyclogher road, to the left you will see a signpost pointing to what is known as Prince Connell's Grave, which lies 50 m off the road. It's a rectangular passage-grave site surrounded by stones. The grave itself has an unusual headstone with an opening at the base. It is said to date to between 2000 and 1500 BC.

At the first opportunity, turn right on to a narrow country lane that leads across the head of the lake to the B52, crossing the border in the process. This was one of many unapproved roads that were blocked off during the Troubles. The diminutive Lattone Lough lies to your left as you turn right on to the B52. Although the traffic is light, cars travel down this road at high speed, and you may consider it safer to take the more scenic route to Belcoo, on an elevated and much quieter road that runs broadly parallel to the B52. While this would appear to fly in the face of the guiding principle of this book – to maintain close contact with the lakeshore – in this instance the higher road provides the better views. For a good deal of the length of the B52, the view of the lake is partially blocked by trees and high hedges.

St Patrick's holy well

It would be hard to match the appeal of the southern shore of the lake, and the cycle along its northern shore offers little in the way of attractions or points of interest until you come to St Patrick's holy well, on the outskirts of Belcoo. The well site that you see today has been much altered and modernised. The site is said to have hosted pagan festivals centring around Lughnasa, the festival of the Celtic god of the harvest, Lugh, before it was christianised by St Patrick. The water is reputed to be the coolest in Ireland – but if truth be told it does not compare with the temperature of the mountain streams you encounter around the shores of Lough Melvin, for example! A cure for nervous conditions and stomach ailments is promised to those who visit the well and avail of its healing properties.

Lying across the road from the well is Templerushin Church and graveyard. Templerushin means 'Church of the Little Wood'. The church is said to have been built around the ninth or tenth centuries and was associated with the well. It has one unusual feature: an opening that can be seen from the altar, known as the leper's window, where lepers could attend services and receive Holy Communion without mixing with the general population.

There is another well to be found in Belcoo. This one is located indoors at the Customs Bar and Restaurant, which looks on to the village's Cottage Lawn. The bar was built on what used to be the premises of the Customs office; the well was discovered when the building was being refurbished. The well, which was preserved and sealed, can be viewed through the floor of the lounge.

Lower Lough Macnean

(25 kilometres)

One disappointing feature of the circuit around the lower lough is that for a significant part of the journey the lake remains out of sight. The consolation is that the circuit is largely undertaken on very quiet country roads and lanes, some of which follow the route of the dismantled Enniskillen–Sligo railway line. Using either Belcoo or Blacklion as your base, travel east, following the path of both the Ulster Way and the Kingfisher Cycling Route. (If you have started from Blacklion, you will have crossed the border into Northern Ireland early on this road.) Along the route, you will catch a view of Inishee Island, one of the lower lough's two islands. Marlbank Nature Reserve lies to the right along with a landmark you cannot help but notice, the Hanging Rock. At the far end of the reserve there is a large stone boulder which is known locally as 'the Saltman'. It is said that this boulder was dislodged during a violent storm and crushed a travelling saltman who used to trade in the area, and who was never seen again after that night.

At the first opportunity, turn left and head down a long straight road which follows the course of the Arney river, crossing one of its tributaries, the Cladagh river, on the way. The Ordnance Survey map for the area indicates that the Ulster Way follows the bank of the Cladagh, making for a crossing of the Arney close to the lakeshore. This route was closed in recent years for insurance reasons. As a result, those wishing to cross the Arney now have to make the 6 km journey to Brockagh Bridge. Having crossed this bridge, turn left, and at the next crossroads turn left again for the westward journey back to the lake. The road follows the route of the dismantled railway line, as mentioned earlier. As you approach the lake, you will see Cushrush Island, also known as 'Rabbit Island'. You should take care not to trespass on what is now private property, as the land on which the railway stood has been sold off to private interests: watch out for a concrete paved lane to your right, which will lead away from the lake up to the busy A4.

Mullycovet Mill

On your way back towards Belcoo, you will pass the sign for Mullycovet Mill, a water-powered corn mill which is located 650 m up the hill to the right of the road. To visit this mill, also known as 'Cleggan Mill', you will either have to leave your bicycle locked at the gate or haul it up the steep climb. The mill, which dates back to the eighteenth century, belonged to the Earls of Erne. In its early days, it had a guaranteed flow of business under a system known as the 'milling soke', whereby the miller was given exclusive rights to process all the grain produced in a particular area. In the case of Mullycovet Mill, this covered all the land owned by the Earls of Erne, whose tenants were therefore obliged to use the mill. The establishment of large milling companies sounded the death knell for this restrictive feudal system. It is thought that the mill operated until the late nineteenth century, after which time it was abandoned. The mill was restored to working order by local interests in recent years. Apart from the mill, there is a grain-drying kiln, and the miller's house is sited in a derelict farmyard nearby to the north.

Just before you reach Belcoo, there is a viewing area underneath Aughrim Hill where you can enjoy a delightful view across to the distinctive table-top profile of the Cuilcagh Mountains.

County Sligo

Lough Arrow

Lough Gill

Lough Gara

W. B. Yeats described Sligo as 'the Land of Heart's Desire'. Like many visitors to the county, he was captivated by the charm and beauty of its lakes, in particular Lough Gill. It is only when you stand on the shore of one of these lakes that you will fully appreciate how they evoked in him such powerful language.

The county's name is derived from the Irish word sligeach, which translates as 'a shelly place'. Its landscape is varied, with an attractive mixture of mountain, rugged coastline, lakes and rivers. It is a haven for archaeologists, with more than five thousand sites of archaeological significance having been identified already, and is an area full of history and fable, featuring both the last resting place for Queen Maeve of Connacht and the location where the legendary Irish warrior Diarmuid met his end. In fact, there is so much of interest in the vicinity of the county's lakes that you will find that the circuits around them will take much longer than usual – which is no hardship.

Sligo is not the sole owner of any of its principal lakes. Even Lough Gill, which lies so close to the county's principal town (recently designated a city) and which has become synonymous with the county, is shared – with County Leitrim. The three lakes included in this section are those where the predominant part of the lake's area lies within Sligo's county boundaries.

Unshin River

Ballindoon

Castlebaldwin

Annaghgowla Isd

Inishmore

Ballynary

Inishbeg

Loughbrick
Bay

Hollybrook
Demesne

Bricklieve
Mountains

Muck Isd

Drumdoe

Ballinafad

Lough Arrow

(25 kilometres)

LOCATION
To the south of County Sligo, 24 km south-east of Sligo town and 6.5 km north-west of Boyle, County Roscommon. A small corner of its southern shore lies in County Roscommon.

LENGTH
7 km/4.3 miles

WIDTH
3 km/1.9 miles at its widest point

AREA
1,250 hectares/12.5 km^2

PUBLIC ACCESS
Brick pier on the eastern shore, Ballinafad pier on the southern shore and Rinn Bán pier on the western shore

MAP
Ordnance Survey of Ireland Discovery Series Map 25, covering Sligo, Leitrim and Roscommon (ISBN 978-1-901496-24-6) and Map 33, covering Leitrim, Longford, Roscommon and Sligo (ISBN 978-1-901496-05-5)

Lough Arrow is a popular limestone water fishery for brown trout. Angling guides point to the fact that it was on this lake that the technique of gnat fishing with dry flies was pioneered in the early 1900s. The mountainous and hilly terrain that dominates the surrounding land is studded with sites of significant historical interest – sites which will be highlighted as the circuit proceeds. Despite the lake's small size, several substantial islands are to be found in the northern half, giving the lake an interesting and attractive profile when viewed from higher ground. The lake is mainly spring-fed, and much of it has no defined shoreline, with fields and woodland melting into the water. It is noted for its boggy foundations, with a soft sandy bottom that quickly gives way, and it can be treacherous for those who might attempt to swim in it.

The circuit around the lake is short but packed with areas of interest. Despite the hilly terrain (mentioned above), the route near the lakeshore is relatively flat. With the exception of a 3 km stretch on the western side, where the N4 has to be used, the loop follows well-surfaced country lanes that see very little traffic. Even on the N4 there is a good hard shoulder, which affords plenty of room for the cyclist, making the overall circuit a safe and pleasant cycle route.

My suggestion for a starting point is Brick pier at Loughbrick bay, halfway up the western shoreline, where there is good off-road car parking. In addition, the nearby Rock View Hotel serves bar food, which will no doubt be welcome after the circuit. Just past the pier there is a trail to the left, which leads down the length of the extensive Annaghloy promontory and is a pleasant extension to the circuit – which you might take either before embarking on the circuit or just after completing it. Moving northwards, you get a good view, firstly of Inishbeg Island, and then its larger counterpart, Inishmore Island. In keeping with tradition, these islands are also known locally as Flynn's Island and Gildea's Island, respectively, after local landowning families.

Ballindoon Priory

A short cycle further on are the picturesque – and easily accessible – ruins of the sixteenth-century Dominican Ballindoon Priory. The priory was established by the Mac Donagh family in 1507 and was one of the last medieval religious foundations to be built. (Of course, the dissolution of the monasteries in Ireland began in around 1536, on the orders of Henry VIII, so Ballindoon Priory was not destined to last long.)

Heapstown cairn

At Ballindoon crossroads, a left turn allows you to continue following the course of the lakeshore, to another junction known as Heapstown crossroads. While you will be turning left to continue across the northern shore, a small off-route diversion across the junction is worthwhile, to view Heapstown cairn, one of the largest cairns to be found outside the Boyne valley in County Meath. The cairn is located on the right-hand side, less than half a kilometre up the road. Unfortunately, it is less impressive than it used to be, as some of its material was taken for road building and its pillar stone is missing. It is thought to be a passage tomb and is regarded as the most important unopened cairn in the Moytura area, the hilly region which runs along the eastern shore of Lough Arrow and throughout which are to be found numerous megalithic tombs and mounds. The Heapstown cairn may have been associated with the Carrowkeel passage tombs found on the slopes of the Bricklieve Mountains – which we will come to later. The cairn is sometimes mentioned as the grave of Ailil, brother of the King of Tara, Niall of the Nine Hostages, a fifth-century Irish chieftain who became high king of Ireland.

The legendary battles of Magh Tuireadh

The Moytura area mentioned above has huge importance in Celtic mythology, although there is some confusion surrounding it. Its Irish name is Magh Tuireadh, which translates as 'the Plain of the Pillars'. The great antiquarian George Petrie visited the area in 1837 and claimed to have discovered the two towers, or pillars, which had given the area its name. He also noted at the time that he had seen two similar towers in an area near Cong in County Mayo. According to Petrie, the two towers measured over 30 m in diameter. The area near Lough Arrow extending to the strand at Ballysadare is accepted as the location for the Second Battle of Magh Tuireadh, one of two defining conflicts for the Tuatha Dé Danann, the divine people of Irish tradition. However, an earlier battle with a similar title – the First Battle of Magh Tuireadh – is sometimes said to have taken place much further south, near the shores of Lough Mask and Lough Corrib in County

Mayo (see page 277). That battle saw the defeat of an established Irish tribe called the Firbolgs, but there is some debate as to its actual location. In some texts, both battles took place in the same area. However, in his book Lough Corrib: Its Shores and Islands, Sir William Wilde (1815–76) claims that the first battle took place on the plain of Moytura, which extends west from the shores of Lough Mask. He describes vestiges of many monuments that are said to mark various defining events that took place on the battlefield over the four days of its duration. To lend credence to his theory, one of these is called 'Eochy's cairn' (pronounced Yeo'hee), a massive tumulus said to mark the spot where King Eochy, the last of the Firbolg kings of Erin, perished after slaying three sons of one of the Dé Danann leaders. Most authoritative sources now discount Wilde's theory in relation to the location of the first battle. However, there is no debate as to the location of the second battle, which resulted in a Dé Danann victory over demons called Fomhoire and the expulsion of the latter from Ireland. It is understandable, therefore, that the area should also be a repository of sites rich in archaeological significance.

Castlebaldwin Castle

Resuming the route across the northern shore, there are appealing views of Annaghgowla Island, also known as 'Lyttle's Island'. If you stop at Bellarush Bridge, which spans the Unshin river, and look towards the lake, you will note a stone structure on a small island. This was part of an eel trap built by the Cooper family, who owned the fishing rights to the lake in the nineteenth century and whose family home, Markree Castle, lies to the north near Collooney. The Unshin drains Lough Arrow to the north, eventually merging with the Ballysadare river before entering the sea at Ballysadare bay. Bear left at the next junction and continue on the tight country lane that skirts the small village of Castlebaldwin. To your right are the ruins of Castlebaldwin Castle – a grandiose title for what is in fact no more than a medium-sized house with high gable walls. This structure, a semi-fortified house, was built at the start of the seventeenth century. Its fortifications consisted of small apertures located in the walls through which muskets could be fired, and a machicolation (a horizontal projection with openings built out from a castle wall, supported by brackets) above the entrance door, from which guns could be fired or stones or boiling oil dropped on attackers.

Carrowkeel passage tombs

At the end of the country lane, you will have to join the busy N4 for approximately 3 km, passing Hollybrook House and Demesne, which obscures the view of the lake. Leave the N4 where it is signposted for the Lough Arrow Scenic Drive. There is a public shore-access facility at Rinn Bán pier, close to this junction. As you make your way towards Ballinafad and the lake's southern shores, the Bricklieve Mountains dominate the landscape to the west. These limestone uplands are host to a series of fourteen cairns, known as the Carrowkeel passage tombs, which date back to between 3800 and 3400 BC. In contrast to the more famous passage tombs in the Boyne valley, the tombs are not decorated. Nonetheless, they are well worth a visit – but you will have to leave your bike behind. They can be accessed using waymarked trails that you can pick up west of Ballinafad. Not only will you see the tombs, but you will also be rewarded by magnificent views across Lough Arrow.

Ballinafad Castle

Ballinafad occupies a pivotal position between the Bricklieve and Curlew Mountains. The village name is an anglicised version of the Irish Béal an Átha Fada, which translates as 'the Mouth of the Long Ford'. With its strategic location, it is understandable that you will find on the western side of the village, close to the N4, the ruins of what must have been a good example of the romantic ideal of a castle. Known as 'the Castle of the Curlews', it was built between 1590 and 1610 by Captain St John Barbe, who was granted the surrounding lands by James I of England. With a round tower at each corner, it resembles a scaled-down version of some of the larger Norman castles built in the early Middle Ages, such as Roscommon Castle or Carlow Castle. It did not last long, however: it was forced to surrender in 1641 when its occupants ran out of water, and by the end of the seventeenth century it had fallen into disrepair.

Drumdoe House and Demesne

Initially, the route across the southern shore from Ballinafad is very appealing. Once you pass Ballinafad pier, the final public shore-access facility on the circuit, there are good views along the wooded shoreline of McDonagh's Shore and Jack's bay, and northwards across the lake. The quiet country lane takes you away from the lake, crossing the county boundary into Roscommon and passing the well-maintained stone walls of Drumdoe House and Demesne. This property, extending to 80 hectares, was owned in the nineteenth century by Judge Flanagan, a local district court judge. After Judge Flanagan's time, the house was bought by Captain Percy McDermott, who also owned land near Coolevin. A local man, Dan Kelly, told me that when he was fourteen years old he used to collect milk for the creamery on a horse and cart to help support his family. The milk of Drumdoe demesne was included on his rounds. One morning, Captain McDermott asked him to bring some apples to the creamery on the cart. Dan refused, and Captain McDermott was apoplectic: he was not used to being turned down but never held it against Dan. On Captain McDermott's death, the demesne was bought by a German family with the surname Praeger, who also own Hollybrook House on the lake's western shore.

Chilean connections

You remain within the frontiers of County Roscommon for only about 5 km, but for the time you are there you lose sight of the lake. Two left turns at the next two junctions will take you on a northerly course, cycling back towards Brick pier. Just before you reach your destination, there is an interesting memorial on the right-hand side in front of the Lough Arrow Touring Caravan Park. It is dedicated to Ambrose O'Higgins, who is described on the memorial as 'the Marquis of Osorno and Baron of Ballenary'. The name of the townland in this area is Ballenary – although it is sometimes listed as 'Ballynary'. O'Higgins, who was born here in 1720, went on to become Governor of Chile and Viceroy of Peru. He died in Peru in 1801. The memorial notes that he was the father of the liberator of Chile, Captain General Bernardo O'Higgins, who was born in Chile in 1778 and, like his father, died in Peru, in 1842.

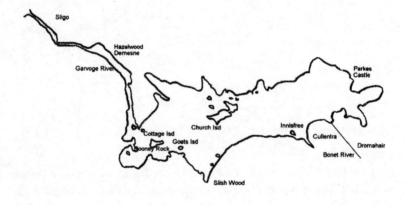

Lough Gill

(49 kilometres)

LOCATION
5 km east of Sligo town. A small corner of its eastern shore lies in County Leitrim.

LENGTH
10.5 km/6.5 miles

WIDTH
3.2 km/2 miles at its widest point

AREA
1,320 hectares/14 km^2

PUBLIC ACCESS
Dooney Rock, Cleaveragh demesne on the banks of the Garvogue river, Half Moon bay at Hazelwood demesne, Sriff bay, Parkes Castle at Fivemilebourne, Cullentra Wood overlooking Inishfree, and Slish Wood

MAP
Ordnance Survey of Ireland Discovery Series Map 25, covering Sligo, Leitrim and Roscommon (ISBN 978-1-901496-24-6)

Lough Gill is an enchanting lake with numerous attractive features. To those of a literary disposition, it will be visited to experience what inspired W. B. Yeats to write such memorable verse. To anglers, it is well known as a salmon and brown-trout fishery. To geologists, it marks a junction between the older metamorphic rock of the Ox Mountains and the younger carboniferous limestone deposits of north Leitrim. To naturalists, the islands and the area around the lake are specially noted for the presence of rare plant species. To archaeologists, the stone-laden hills offer a wealth of significant sites abounding in crannógs and megalithic cemeteries. To hillwalkers, it offers trails of mixed character and alluring views. To cyclists, it presents a route that is full of variety, and the opportunity to experience several off-road sections that range from easy to challenging.

Legendary origins

The lake is said to be named after Gile, the daughter of an ancient Irish warrior called Romra. Another warrior called Omra, sometimes reported to be Romra's brother, fell passionately in love with Gile after seeing her bathing, but the match did not find favour with her father. Romra found the two together and was so incensed that he immediately challenged Omra in battle. The ensuing fight was so brutal that both men were fatally wounded. Gile had lost the two men she loved most dearly, and her tears of grief were so abundant that they formed the lake that was named after her. The graves of the two warriors are marked by two cairns lying close together at Carns Hill Forest Park, overlooking Lough Gill on its western shore.

Description

The lake has a complement of twenty-two islands, the largest of which is Church Island, located roughly in the centre of the lake, and the most famous being Inishfree, close to the southern shore. Most of the islands are concentrated in the western half of the lake. Its deepest point has been measured at 35 m. The waters of the lake are often depicted as being blue, unlike the waters of most other freshwater lakes – as shown in a well-known painting by W. H. Bartlett. Perhaps this gives an indication of the true origin of its name, as the translation of its Irish name, 'Loch Gile', is 'bright lake'. On a cloudless day, the lake waters certainly reflect the blue skies, creating a breathtaking scene.

Sligo Abbey

A good starting point for the loop around the lake is the car park at Cleaveragh demesne, on the banks of the Garvogue river, Lough Gill's outlet to the sea. The car park, which is near Sligo racecourse, can be easily accessed off the N4 by following the signs for either the racecourse or the regional sports centre located close to the car park. To ease you into the circuit, a well-surfaced 1 km-long path leads towards Sligo town, along the banks of the river to a pedestrian bridge that allows you to cross the river without going into the town centre. One building in the town – Sligo Abbey, the town's only surviving medieval structure – has a connection with the lake, and it is certainly worth a visit before you continue on the circuit. It is located very close to the riverbank on the left-hand side before you come to the first road-bridge on the route. It was originally built in 1253 for the Dominican friars but had to be rebuilt in 1414 after it was accidentally burned down; as a result, the style of the present ruins are more representative of fifteenth-century Gothic architecture. Its connection with the lake relates to its silver bell, which is said to lie at the bottom of Lough Gill. It is said that only those who are perfect can hear it on the occasions when it peals.

Hazelwood demesne

Using the R286 to exit Sligo on the north side of the river, you will come to a right turn that leads to a route extension that I recommend should be included in any loop of Lough Gill. This extension will add roughly 9 km to your overall journey. The road leads to Hazelwood demesne, a Coillte woodland amenity on the grounds of an estate that used to be the ancestral seat of the Wynne family. Hazelwood is the place that Yeats referred to in his 'Song of the Wandering Angus'. To access the demesne, you should take the first turn left, roughly 1 km down the road; this will lead you through a delightful wood-shaded road to the aptly named Half Moon bay. From the car park, there is a trail that leads along the lakeshore and the eastern bank of the Garvogue river. Along the trail, you will find a series of large sculptures, including several with odd names such as 'Fergus Rules the Brazen Cars'. Unfortunately, as of the time of writing it is not possible to visit Hazelwood House, at the centre of the demesne. The house was occupied by the Irish subsidiary of a Korean company until April 2006, at which point it was bought by a consortium of local business people who hope to develop it as a tourism facility. The erstwhile residents, the Wynne family, were once substantial landowners in County Sligo. The house was built in 1730/31 for Lieutenant General Owen Wynne to the design of the German architect Richard Cassels, who also designed Leinster House in Dublin.

Returning to the main road, there is a steady climb as you leave Sligo town behind. After a couple of kilometres, there is a right turn signposted for St Angela's College. This is known as the Loop Road, and the climb up the road is fairly steep. Once you reach the top of Loop Road, you are rewarded with a superb view across the western end of the lake and back towards Sligo.

The entrance to St Angela's College is on your right as you continue on Loop Road. This third-level institution occupies another former residence of the Wynne family. A little further on, a turn to the right leads on to a narrow lane – and some more climbing – before you embark on a hair-raising descent, on a bumpy surface, to rejoin the R286.

Parkes Castle

For a while, the road runs along the shoreline. You then cross the Sligo/Leitrim county boundary before reaching the impressive Parkes Castle. This is described in the tourist literature as a fine example of a fortified manor house of the seventeenth-century plantation period. It comprises a manor house surrounded by a bawn, or enclosed yard, with towers on the angles, and it occupies a romantic setting on the shores of the lake. The restored fortress is in fact a hybrid construction, resulting from the architectural marriage of the original stronghold of the O'Rourkes, rulers of east Breffni (County Leitrim on today's maps) in the twelfth century, and the improvements made by Robert Parke in the seventeenth century. It is Parke's construction that dominates the site today. The last O'Rourke to occupy the castle was Brian na Murtha O'Rourke, who was executed in London by Queen Elizabeth I in 1591 for refusing to pledge allegiance to her. It is said that he asked to be hanged not by a rope but by a 'wythe', or twig, when he was executed at Tyburn.

When the site was being excavated in the 1980s, archaeologists discovered the foundations of a rectangular tower house dating back to the sixteenth century. The tower house would have been occupied by Brian na Murtha O'Rourke prior to his being summoned to London. Following his execution, his lands fell within the ambit of a plantation scheme devised by James I, whereby the lands of Irish leaders who had fled the country in 1607 were granted to English and Scottish settlers. It was under this scheme that, in 1620, the castle became the property of Captain Robert Parke, who set about rebuilding it to suit his own requirements. At the time, it was known as Newtown Castle: you will see this name beside the coat of arms over the entrance.

The Parkes came from Kent in England. In 1642, Robert Parke was imprisoned on suspicion of being disloyal to Cromwell. The castle was seized by a Catholic family, the O'Haras, who held it until 1652, when they surrendered it to Sir Charles Coote, who was in command of Cromwell's troops. When Charles II was restored to the throne in 1660, the castle was returned to Parke, who later became Member of Parliament for Leitrim. The Parkes' only surviving daughter, Anne, inherited the castle on the death of her parents. She later married Sir Francis Gore, an ancestor of the Gore-Booths of Lissadell. The last resident of the castle was their son Robert Gore.

The castle, a designated heritage site, is open to the public and is well worth exploring before you move on. The Rose of Innisfree tour boat, which moors adjacent to Parkes Castle, offers daily trips around the lake, passing by the Isle of Inishfree.

Dromahair

The Leitrim end of Lough Gill is extremely picturesque, and you get full value, as the road runs along the shoreline for most of the journey until it pulls away to the south towards Dromahair. This village is a good place to stop for refreshments, either at the Abbey Manor Hotel or the Blue Devon pub. As you enter the village, there is a turn to the right up to a nineteenth-century mill which is now a storage building for the Office of Public Works and is not always accessible. The building houses three brass cannons salvaged from the Spanish Armada shipwrecks at Sreedagh beach north of Sligo, and also a famine soup pot from the workhouse at Manorhamilton.

At Dromahair, you join the route of the Sligo Way, going down a narrow alleyway past the side of the Abbey Manor Hotel and across the River Bonet. You pass the front door of a small cottage in order to reach Creevelea Abbey, a Franciscan friary founded by Margaret, wife of Owen O'Rourke, a local chieftain of the O'Rourkes of Breffni, in 1508. Established a year after Ballindoon Priory, mentioned in the Lough Arrow circuit, this was the last such foundation before the suppression of the monasteries. After the abbey, you can follow the waymarked trail through a combination of pleasant country lanes, rougher ground and woodland to Cullentra Wood. These predominantly oak woods have been in existence for more than three hundred years and have recently been augmented by tree-planting associated with the People's Millennium Forests Scheme. The name 'Cullentra' is derived from the Irish 'Cullentragh' or 'Culennagh', meaning 'a place abounding in holly'. It is part of a larger area of woodland known as Slish Wood, which you will come to later.

'The Lake Isle of Inisfree'

The wood is located on a headland almost directly overlooking the island of Inishfree (the Irish for 'heather island') – an island mainly associated, of course, with W. B. Yeats, who wrote his celebrated poem about the island in 1890 while he was living in London. Although Yeats was born in Dublin, Sligo was his spiritual home, and the area around Lough Gill features frequently in his poetry. His connection with Sligo came through his mother, who was born Susan Pollexfen, the daughter of a wealthy shipping merchant and mill owner. Yeats used to spend much of his summer holidays in Sligo and built up such an enduring affection for the county that he expressed the wish in his poem, 'Under Ben Bulben', to be buried there. He died in Roquebrune in France in January 1939, but it was not until 1948 that his wish was fulfilled and his remains were interred at Drumcliffe Churchyard, to the north of Sligo, where his great-grandfather used to be rector.

The viewing area at Cullentra provides some of the best views of the lake and its hinterland. Apart from Inishfree, there are panoramic views across to the northern shore, where you can make out Parkes Castle and St Angela's College. Above them is a range of hills known as 'the Sleeping Warrior'. Its profile resembles the side view of a burial casket you might see in a church, with head, body and feet clearly delineated. In the distance is the distinct table-top and cliff-edged profile of Ben Bulben, which rises 527 m. The mountain is named after a son of Niall of the Nine Hostages.

Slish Wood

The Sligo Way continues on a rising boggy path on the lower reaches of Killerry Mountain, maintaining close contact with the lakeshore. This continues for more than 1 km and is not suitable for bicycles, although the more adventurous and well-equipped mountain biker might be tempted to give it a go. The alternative is to backtrack through the country lanes to the R287 and pass Killerry Mountain on its southern side. My preference is to haul the bike over the hill along the boggy path until you reach a small stream which marks the rear entrance to Slish Wood – with its firmer forest trails, which at present are only open to walkers. The journey to the stream following the marked route takes about forty-five minutes: if you were not to choose it you would miss out on delightful views and the quiet trails of Slish Wood.

Dooney Rock

It is a short journey from the exit of Slish Wood to another place that struck a chord with Yeats, Dooney Rock. The name features in another of Yeats' poems, 'The Fiddler of Dooney'. The rocky outcrop can be accessed through the car park, but if you want to enjoy the views from the top you will have to secure your bicycle in the woodland that surrounds it, as there is a very steep climb to the summit. You will be rewarded with a good view of the lake in its entirety, and further afield.

Two young botanists lost their lives in the lake area between Dooney Rock and nearby Goat's Island on 4 August 1883. Thomas Hughes Corry and Charles Dixon were surveying the lake and its islands for a botanical report when they drowned near the shore.

Tobernalt holy well

Another site of interest is Tobernalt holy well, which lies a short distance away on the small road that runs along the western shoreline of the lake. The site, reputed to be where St Patrick said Mass, is more elaborate than that of most holy wells you will encounter. Beside the well there is a mass rock, in which there are five indentations, said to be the saint's fingers. It is said that if you suffer from backache, you should rest your back against this rock and your pain will disappear, and the water from the well is said to cure headaches and poor eyesight. There is a regular flow of visitors who collect water to be used for various religious purposes, including watering flowers and plants on graves. The site is very well maintained, with modern Stations of the Cross around the rock garden that surrounds the well. A Mass is celebrated at the well every year on Garland Sunday, the last Sunday of July, which marks the beginning of Lughnasa, the festival of the Celtic god of the harvest, Lugh.

A steady climb away from Tobernalt along a tree-shrouded road leads back towards Cleaveragh, completing one of the most interesting loops you will undertake around any lake in Ireland.

Moygara Castle

Mullaghatee
Mountain

Crow
Isd

St. Patrick's
Rock

Boyle River

Cuppanagh
Bridge

Inch Isd

Derrymore
Isd

Monasteraden

Clogher
Stone Fort

Edmondstown

Lung River

Breedoge
River

Ballaghaderreen

Breedoge

Lough Gara

(41 kilometres)

LOCATION
In the southern reaches of County Sligo

LENGTH
4 km/6.5 miles

WIDTH
4 km/2.5 miles at its widest point

AREA
1,100 hectares/11 km^2

PUBLIC ACCESS
Cloonloo pier at Cuppanagh on the northern shore, and Clooncunny Bridge on the eastern shore

MAP
Ordnance Survey of Ireland Discovery Series Map 32, covering Mayo, Roscommon and Sligo (ISBN 978-1-903974-24-7), and Map 33, covering Leitrim, Longford, Roscommon and Sligo (ISBN 978-1-901496-05-5)

Some locals speak ruefully of Lough Gara's better days, in the time before the lake was drained in 1952 to relieve winter flooding in the area. They say that the character and profile of the lake changed for the worse and have described the drainage scheme as amounting to the rape of the lake. In a locally produced publication, it was described as once having been 'a deep attractive lake teeming with fish, wildlife and boats'. The publication adds that its summer level 'can now be measured in inches rather than feet'.

To the outsider, gazing upon it for the first time, the lake, which has been designated as a wetland wildfowl reserve, still appears beautiful, framed by bogland and meadow. As a cycling circuit, it ranks highly not only because of its aesthetic appeal but also because the surrounding roads are extremely quiet and relatively flat, with only two moderate ascents. In addition, there is a lot of interesting history to be discovered, relating to both the lake itself and its hinterland.

Description

Lough Gara has three distinct elements – 'the callow', or shallow parts, to the south, a middle section known as 'the upper' and a northern section called 'the lower'. Most of its surface area lies in County Sligo but the callow stretches into County Roscommon. It is the most westerly reach of the River Shannon system and feeds into that great river via the River Boyle and Lough Key, both to the east. Its waters are replenished by the Lung river, which empties its flow from the southwest, and the Breedoge river, which discharges its waters from a southeasterly direction. The lake has a number of large islands but some of these are more like peninsulas, with connections to the mainland during the summer months. It also abounds with man-made islands or crannógs, many of which were revealed by the drainage of the 1950s.

The lake's name is derived from its connection with the O'Gara clan, who ruled this territory for about four hundred years up to the middle of the seventeenth century. Before their time, it was known as Lough Techet. The O'Gara chieftains occupied Moygara Castle, which is sited on a majestic and strategically imposing position overlooking the lake's northern shore.

Impact of drainage schemes

The drainage scheme referred to above was not the first attempt to relieve the problems of winter flooding in this generally low-lying area – a previous effort was made in 1859 – but it was the severity of the 1952 scheme that has generated so much debate. Initially, the plan was to reduce the water level by 1.22 m, to bring it in line with the summer levels. However, complications set in when an abundance of crannógs and other artefacts, which had been hidden from view for many years, began to be revealed. In order to facilitate excavation of these finds, it was decided to lower the water level by a further 1.2 m. It is reported that as many as 360 crannóg sites were discovered, as well as canoes and other evidence of significant human habitation dating back over two thousand years. Some of these artefacts are now on display in the National Museum in Dublin. I share the view, held locally, that it might have been better if the finds could have been housed in a purpose-built museum in the area that would help to celebrate the importance of this lake and its hinterland in its original archaeological context.

There has been a lively academic debate among archaeologists and limnologists, specialists in the physical and geographical phenomena of lakes, about what constitutes a crannóg. The simplest and most widely accepted definition is that the term applies to a man-made island. The word is an Irish term derived from crann, meaning 'tree', as timber was used as the basis for the island. According to Christina Frederengen, a Swedish archaeologist who carried out detailed excavation of some of Lough Gara's crannógs, the term first appeared in the thirteenth century, in an entry made in the Annals of Loch Cé.

Historical research suggests that crannógs, which are usually circular or oval in shape, were probably aristocratic dwellings used from the sixth to twelfth centuries. In some cases, they were natural circular islands that were modified using layers of material such as stone, peat or brushwood, surrounded by a retaining ring of timber piles or a pallisade. In other cases, they were completely artificial, built up using these materials. Mainly defensive in nature, they were generally built close to the shore and would have been reached by boat or causeway. In many cases, the causeway would have been submerged, keeping the crannóg safe from those who did not know how to access it. In later years, they were used as areas where livestock could be isolated or protected from predators such as wolves or humans.

Approximately two thousand crannógs have been identified on the island of Ireland, with most found in smaller lakes or in the bays of larger lakes. They are concentrated in a belt stretching east to west from Down through Cavan and Monaghan to Sligo, Leitrim and Mayo. There are very few in the south of the country. The number of cran-nógs in a lake varies substantially: while Lough Gara has many, none exist on some of the larger lakes included in this book. It is thought that the dangers associated with open waters militated against cran-nógs being sited in the larger lakes. Crannógs are used to this day as bird hides and mooring jetties, and in some cases even as sites for illic-it distilleries.

Monasteraden and St Attracta's well

The village of Monasteraden is a good starting point for the lake cir-cuit, as you can retire to Drury's bar for refreshment after your exer-tions. The village, located 6 km south of the Gurteen–Boyle road (the R294), takes its name from a monastery founded by St Aidan, who is believed to have flourished in the sixth century. The ruins of the monastery can be seen in the village cemetery. To the south of the vil-lage, on the road that eventually leads to Ballaghdereen, is St Attracta's holy well, whose surrounds were restored in recent years. St Attracta, a descendant of Coelbadius, an Irish king of the early fifth century, is the patron saint of the area. It is said that she met up with St Patrick nearby and that he consecrated her to God. Afterwards she established a nunnery at Killaraght, on the eastern shore of the lake. On her feast day, 11 August, locals gather at the well for prayers. On a flagstone to the rear of the well is a beautiful stone carving of the crucifixion.

The Monasteraden Monument

On the road that leads north from the village towards Moygara Castle there is a monument known as 'the Monasteraden Monument', erect-ed in the memory of two men who died in tragic circumstances in 1881. The story behind their deaths reveals the passions roused in rela-tion to the land – passions that led to the creation of the Land League movement. The story goes that there was a process server called James Broder on his way to the area to serve eviction notices on several ten-ants of the estate of Major Arthur French. He was escorted by four policemen, led by a Constable Walter Armstrong. Where the monu-ment now stands, the policemen were confronted by a gathering of people, including women and children, who blocked their progress. A week earlier, there had been an altercation between the process server and a man called Brian Flannery, who was now standing in the midst of the group. During that altercation, the process server received a facial wound, and he swore vengeance. In the later event, the situation soon became heated, and the process server drew his gun and shot Brian Flannery dead. Constable Armstrong, who had earlier warned the crowd that he would give orders to shoot into them if he and his colleagues were not allowed to pass unharmed, then shot another member of the group called Thomas Corcoran. The crowd was incensed and rushed the process server and the policemen. Such was their rage that they killed Armstrong on the spot. They would no doubt have meted out the same punishment to the others had they not been thwarted by the escape of three of them and the courageous action of a young girl, who threw herself upon one of the policemen and pleaded for his life. A number of people were eventually arrested and tried for the killing of Constable Armstrong, but they were acquitted, on the grounds that the constable had no authority to shoot into the crowd because he had not first read them the Riot Act.

On the centenary of the men's death, in 1981, the monument in memory of the two men, Brian Flannery and Thomas Corcoran (which had been erected in 1917 by locals), was restored, and five hundred people attended a ceremony there.

Clogher stone fort

As you head south out of Monasteraden in the direction of Ballaghdereen, a signpost points to an opening in the stone wall of the demesne that lies to the west – and to Clogher stone fort. A semi-circular railing has been erected on the opposite side of the opening, and this will prevent you from bringing your bicycle any closer. The stone ringfort, also known as a cashel, lies hidden by trees on an elevated position about 100 m to the right of the gate. The magnificently preserved oval stone structure, which was partially rebuilt in the nineteenth century, is reminiscent of a similar prehistoric fort, Dun Chonchubhair, on Inis Mór, the largest of the Aran Islands. The Clogher stone fort uses no mortar, and its walls are fifteen feet thick at ground level and rise about 3 m. Steps are to be found on the inside walls, allowing you to access the top of the wall at various places.

Within the fort are four souterrains, or subterranean chambers, which run under the walls for considerable distances. This is what is known as a bivallate site, in that there are two defensive walls. The remains of a surrounding earthen bank constitute the second wall; use of stone and earth is an unusual combination in ringforts. Although there is no sign of them today, sharp up-ended rocks would have been placed in the ground outside the fort as a defence against approaching enemies. It is a pity that there is no information panel on the site explaining the fort's origins and history.

The southern shoreline

At the Edmondstown crossroads, to be found south of Monasteraden just beyond the county boundary with Roscommon, turn left to start tracking the callow (or southern shore) of the lake, which is largely hidden from view. Indeed, you will be out of sight of the lake for a long period as the trail takes you through very low-lying flat land, with extended views as far as the eye can see. It is easy to understand how these plains can be so easily flooded – and the resulting pressure for the lake to be drained. As you make your way towards the bridge across the Lung river, one of the lake's two principal feeders, you pass by the open paths of the disused railway line, which stretch between the fields on either side of the road. This was a branch line off the main Sligo line that served Ballaghderreen. It opened in 1874 and was closed in 1963. At one stage, eight trains a day passed on this line, with additional 'specials' being run on fair days. Many a local lad (and lass) started the sad journey of emigration to Britain or the United States at the Island Road station near Monasteraden.

After you cross the Lung, open bogland surrounds you on all sides. These are ideal cycling conditions, on flat, well-surfaced, traffic-free roads, in pleasant surroundings. William Bulfin, in his Rambles in Eirinn, remarked that 'the pedals of a bicycle are very faithful and accurate topographers'. He continued that 'if there is a hill to be found on the road at all, they will find it and report it to you at once'. There are no hills here, however, and you will find that you make speedy progress.

At the first opportunity, turn left, passing through a small forested area, before another turn left brings you to the entrance to Roscommon County Council's water-treatment works. Bearing right here will set you on an extended course towards the village of Breedoge, which marks the south-eastern extremity of the lake circuit.

Abandoned homes

Along the trail to Breedoge, you cannot help but notice the number of abandoned and derelict cottages. This is a recurrent feature of this lake circuit. As you will see later, the process of dereliction and abandonment did not respect class: an imposing country house, with walled gardens, at Kingsland fell to the same fate. Some of the cottages and dwellings that are still inhabited are in a shocking state, considering the country's recent economic advances. I came across ramshackle mobile homes that must have been at least thirty or forty years old, and cottages with broken windows and holed roofs where there was evidence of habitation, in the form of smoke swirling up from rickety-looking chimneys.

The trail along the eastern shore of Lough Gara does not yield any view of the lake. After crossing over Breedoge Bridge on the R361, you pass through the village and then turn left just after Kingsland National School. This road continues west, past the abandoned country house mentioned earlier, before turning sharply right, to set you on a northerly course that tracks the eastern shore of the lake. Once again, the haunting presence of abandoned cottages is a constant companion. A signpost pointing to Templeronan burial ground, just after you have crossed back into County Sligo from County Roscommon, is the signal for you to turn left towards the top of the lake. This road tracks the course of the Boyle river to Cuppanagh Bridge, where you are once again reunited with views of the lake.

St Patrick's Rock

The area around Cuppanagh yields some of the best views of the lake at shore level. The burial ground occupies a dominant position close to the shore and is still in use, despite its remote location. Further along, there is a car park and slipway, before, to recall William Bulfin's words, your pedals tell you that you are starting to climb. Just before you have to really dig in to the pedals, there is a recessed sign indicating the location of St Patrick's Rock. According to local legend, St Patrick passed this way on his travels around Ireland, and rested on the rock that now bears his name. The rock is unusually shaped, with an indentation that is perfect for sitting on at one end and a deeper, scooped-out hole at the other.

Moygara Castle

The climb from Cuppanagh to the R294 is steady but not too arduous. There are some excellent views along this road, particularly from the elevated position around Mahanagh bog. As you pass by Maxwell's Brown Trout Inn, you catch the first glimpses of Moygara Castle across the north-eastern extremity of the lake. Not too far ahead is the turn left for Monasteraden. Before the gradient of the road starts to increase, there is a road to the right, just past a stone cottage called 'The Downs', that leads to Moygara Castle. This was home to the O'Garas, lords of Coolavin until 1581, when a force in the service of the governor of Connacht, Captain Malby, attacked and burned the castle, killing Diarmuid Óg Ó Gara, son of the head of the O'Gara clan. Afterwards, the castle fell into disuse, but it must have been occupied at some stage during the seventeenth century, because it is said that the authors of the Annals of the Four Masters were entertained there by their patron, Feargal O'Gara.

You can still appreciate how impressive this castle must have been in its heyday. It occupies a magnificent site overlooking the lake, with Mullaghtee ('the Hill of the Fairies') rising behind it, and views across to Keash Hill and the Curlew Mountains in the distance. The castle and the surrounding curtain walls of its bawn almost form a square. At each corner are rectangular flankers, or towers. One strange feature of the castle, pointed out to me by Frank O'Neill, who is its proud present-day owner, is that the flankers are not sited squarely at the ends of the curtain walls but instead are slightly out of position in relation to the walls. The walls now enclose a grassed area where some of the castle's buildings would have been located. The dwelling area is believed to have been on the west side. In the south-eastern tower is a sycamore tree: according to local legend, this grew from the shoot of another from which the O'Garas used to hang wrongdoers. Another local legend tells of a tunnel that led from the castle to the lakeshore, where there were golden gates at the entrance. Nobody has ever unearthed traces of a tunnel, much less the gates.

Returning to the lakeshore road to Monasteraden, the steady incline yields excellent views of the full expanse of the lake. A turn to the left affords you the opportunity to take on a small loop that descends closer to the lakeshore on a country lane that twice crosses the route of the dismantled railway line. This lane emerges opposite one of the area's best-known landmarks, the Shroofe ball-alley. The alley, believed to have been built more than a hundred years ago, has recently been refurbished and restored to its former glory. A short ride returns you to Monasteraden, and a chance to avail of the hospitality of Drury's pub.

County Roscommon

Lough Key

County Roscommon lies at the heart of Ireland and is the ancient capital of the province of Connacht. Its name is derived from that of St Coman, who is said to have lived in the area in the eighth century and to have established a seat of learning on the location where Roscommon Abbey was later founded by the Dominicans. While it has no access to the sea, two-thirds of the county is bounded by water, with the River Shannon providing the boundary in the east and the River Suck doing so for most of the western part of the county. One third of the county is under bog, and while there is excellent farmland in the centre, the county is dotted with small lakes, making it an angler's paradise. It is also a haven for historians, archaeologists and those interested in myths and legends, with particular focus on Rathcroghan near Tulsk, the burial place of the kings of Ireland and Connacht, where you can find more than twenty ringforts, burial mounds and megalithic tombs. It was from Rathcroghan that Queen Maeve of Connacht embarked on the pursuit of the great brown bull of Cooley, mentioned earlier.

Roscommon is steeped in history and is proud with having provided Ireland with eleven high kings, including the last, Rory O'Conor. The O'Conors of Connacht have historically been the dominant clan in the county and are Europe's oldest family, tracing an ancestral line as far back as 75 AD, to Feredach the Just. Their ancestral home, Clonalis House, is near Castlerea in the west of the county. In more modern times, Roscommon provided Ireland with its first president, Douglas Hyde, who was born at Kilmactranny, where his father was rector, and was christened at Castlerea.

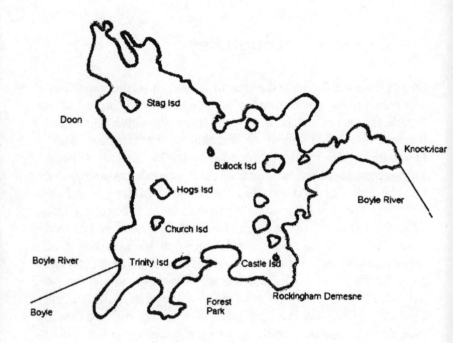

Lough Key

(25 kilometres)

LOCATION
Just under 2 km north-east of Boyle, County Roscommon

LENGTH
4.5 km/2.8 miles

WIDTH
4 km/2.5 miles at its widest point

AREA
900 hectares/9 km^2

PUBLIC ACCESS
Doon Shore to the north-west, Drum Bridge, and through the Forest Park

MAP
Ordnance Survey of Ireland Discovery Series Map 33, covering Leitrim, Longford, Sligo and Roscommon (ISBN 978-1-901496-05-5)

Lough Key is a very popular lake, mainly due to its navigable connection to the River Shannon and the forest park which runs along its southern shoreline and can be easily accessed off the N4. With its proliferation of islands and wooded shoreline, it is widely acknowledged to be one of the most attractive lakes in Ireland. It is best admired in all its splendour from the high vantage point at Doon, off the N4 at the foothills of the Curlew Mountains, which lie to the west. Despite all the visitors and boat traffic, Lough Key is also a popular lake for anglers: the largest pike caught in Ireland, weighing 17.69 kg, was caught here in 1993.

Legendary origins

The lake's name has its origins in Celtic mythology. It is the anglicised version of 'Cé', a druid associated with Nuadha of the Silver Arm, king of the Tuatha Dé Danann. 'Cé' may be an abbreviation of 'Dian Cécht', one of the craft-gods of these people. It was he who replaced Nuadha's arm with a silver one after the king had lost it during the defeat of the Firbolg at the first Battle of Magh Tuireadh, thereby allowing the king to regain his crown. (Physical defects were incompatible with the concept of kingship, and Nuadha had had to step down for a period after being wounded in battle.) However he is described, the druid or craft-god's association with the lake is due to the fact that he was said to have drowned when the waters of the lake burst forth from the earth and engulfed him.

The Annals of Loch Cé

The lake itself, and the surrounding area, are steeped in history, with many battles having been fought on its shores. Some of this history is documented in records such as the Annals of Loch Cé. These combine text written by the O'Mhaoilchonaire family in the fifteenth century, with later writings of sixteenth-century origin scripted by the O'Dhuigheanain family, who had a special relationship with the MacDermot clan, the owners of a fortress on Castle Island near the southern shores of the lake. The Annals were compiled on Castle Island, and it is believed that some of the entries were compiled as late as 1636.

A convenient starting point for the loop around Lough Key is Boyle harbour. This is located at the end of the Boyle Canal and can be reached via the Boyle turn-off from the N4. The harbour is less than 1 km from Boyle town – which should be visited as part of the cycle route because of its interesting history and its numerous historical sites. In addition, there is a quiet road leading from the town directly to Lough Key Forest Park.

Boyle Abbey

The well-preserved ruins of Boyle Abbey greet you on arrival at the town. The abbey was founded by the Cistercians in 1161. The Order was introduced into Ireland in 1142 by Maelmhadhóg Ó Morgair, better known as St Malachy, who persuaded St Bernard of Clairvaux to send a group of his Cistercian monks to help establish a monastery at Mellifont, near Drogheda, County Louth. Its foundation introduced the European way of monastic life, based on austerity and simplicity, to Ireland. At its height, it was the mother-house to twenty-two monasteries, including Boyle Abbey. In 1227, however, there was a split between the Anglo-Norman and the Irish abbeys in Ireland. This resulted in Boyle Abbey affiliating itself directly to Clairvaux rather than Mellifont. Over the centuries, the abbey suffered from various attacks, and its death knell as a religious foundation was sounded with the signing of the Act of Suppression in 1534. Because of its remoteness, it survived somewhat longer than monasteries closer to the east coast, but in 1584 its abbot, Glaisne Ó Cuilleanáin, was executed in Dublin for refusing to renounce his allegiance to Rome. For the following two centuries, the abbey was occupied by military forces and became known as Boyle Castle. In 1892, it was declared a national monument, and it is now managed by the State's Heritage Service.

King House

The route to the town centre leads to another historical building worth visiting, King House, the eighteenth-century home of the earls of Kingston. King House was built in the 1730s for Sir Henry King, a descendant of Sir John King, who had succeeded to the lands of the MacDermots, lords of Moylurg, in 1617. The house, which has three storeys over a basement, was designed by William Halfpenny, an assistant of Sir Edward Lovett Pearce. Sir Henry died in 1739 and the King family did not remain long at King House, moving to Rockingham House on the shores of Lough Key in 1788. The house subsequently became a military barracks for the Connaught Rangers Regiment. It was restored in the 1990s and now houses an interpretive centre focusing on the history of Connacht, the King family and the exploits of the Connaught Rangers. The local tourist office and a coffee shop are housed within the grounds.

Lough Key Forest Park

The road to the south of Boyle Abbey will lead towards a secondary access to Lough Key Forest Park. The road follows a marked historical trail and passes under an archway, providing access to a small bridge across the N4 and leading to the park where there is no entry charge for cyclists. The 324-hectare forest park was previously part of the Rockingham estate, once the property of the MacDermot clan, who, as mentioned earlier, had lost their lands to the King family. Robert King, a grandson of the original grantee, built a house, which he named Rockingham, on the site of one that had been erected by Brian MacDermot a hundred years earlier, and in 1788 the family moved to the lakeshore. The first Rockingham House was destroyed by fire, and in 1810 another Robert King, known as Lord Lorton, replaced it with a mansion designed by John Nash. Although it survived much longer than the first one, this house also succumbed to fire, in 1957, and was never rebuilt. A 24.4 m-high observation tower, called the Moylurg Tower, now stands on the site, and provides magnificent views of the lake and its islands.

An exploration of the grounds reveals numerous interesting attractions – mostly remnants of the Rockingham estate – including a bog garden with stepping stones made of wooden logs, various tunnels, canals and bridges, a wishing chair and a number of gazebos. The canals were used to bring in turf from the local bog. There is also a jetty and mooring facility on the lakeshore, where you can avail of boat tours around the lake and its islands. Moreover, there are around 10 km of forest trails within the park, providing delightful and gentle off-road routes, all of which are open to cyclists.

Trinity Island

Two islands rich in history lie close to the lakeshore at the forest park. The first of these is Trinity Island. The island is in private hands, but you can get a close view of it if you choose to include an off-road extension through the tree-shaded routes to the lakeshore that can be found to the left of the access road. The island possesses the only surviving example of a White Canons monastery in Ireland. This order, also known as the Praemonstratention Canons, was founded by St Norbert at Premontre in France in 1120. Trinity Abbey was established in the early thirteenth century by Clarus Macmailin, archdeacon of Elphin, and the order remained on the island until 1608, when the abbey was confiscated by King James I. The island is also of interest to fans of W. B. Yeats. While Inishfree on Lough Gill captured his heart, the poet was also taken with Trinity Island, and planned to set up a community on the island. His plans did not come to fruition, however. Also to be found on the island are a number of graves, including that of Sir Conyers Clifford, military governor of Connacht, who was killed during the Battle of the Curlews on 15 August 1599. At the time, he was on his way from Athlone to relieve a siege at Collooney Castle, but he was attacked and defeated in a brief battle by a force led by a number of Irish chieftains, including Red Hugh O'Donnell, the MacDermots and the O'Rorkes.

Two other graves are connected with a tragic love story known locally as 'the legend of Úna Bhán'. There are various versions of the legend, but all of them begin with the ill-fated relationship that developed between Úna, the daughter of the head of the MacDermot clan, and a young man of low birth called Tomás Laidir Mac Costello. Úna's father frowned upon the relationship and, in order to prevent the couple from meeting, confined her to the MacDermot castle on nearby Castle Island. In one version of the tale, Úna died from a broken heart due to her confinement and was buried on Trinity Island. After her death, Tomás used to swim out to the island to keep vigil at her grave, but he drowned during one of his crossings. Another version of the story relates that it was Tomás who died first, drowning while attempting to visit Úna on Castle Island, and that she died, grief-stricken, soon afterwards. Both were buried on Trinity Island beneath two trees that intertwined above their graves, forming a lovers' knot.

Castle Island

Castle Island, which you will be able to see from Moylurg Tower, has a nineteenth-century mock castle built on the site of a fifteenth-century tower house that was an important base for the MacDermots. It was also the island on which the Annals of Loch Cé were written, as mentioned earlier.

Knockvicar

Exiting the forest park without returning to the main entrance on the N4 can be a little confusing. However, by heading in an easterly direction and using the trails and roads that run reasonably close to the lakeshore, you will emerge on to the R285 at an unusually shaped house with an hexagonal-shaped second storey. This road leads to the village of Knockvicar. Before entering the village, you cross the Boyle river 300 m south of Clarendon Lock, which is used by cruising boats to access Lough Key. This can be a very busy spot during the height of the cruising season, and you may wish to amuse yourself for a while watching the boats come and go.

A short way along the shore from Knockvicar is The Moorings bar and restaurant, which is popular with the boating fraternity but is also welcoming to cyclists. Its bar counter is over three hundred years old and has seen all types of travellers slaking their thirst. The route to the lake's most northerly point is along a quiet tree-shaded road which offers no access to the shoreline and only intermittent views of the lake. As you turn sharply south, a steady climb along the western shore rewards you with excellent views, and you will begin to appreciate just how many islands there are on this relatively small lake. Just before you reach the N4, there is a road to the left which leads down to the shore near Hog's Island. To the south is the smaller Church Island, where you can see the ruins of a sixth-century church and a later monastic settlement dating back to the twelfth century. If you wish to avail of the best view of the lake in the surrounding area, you will have to backtrack a short way along the N4 to Doon viewing point, which is guarded by an imposing sculpture of a Irish chieftain. The sculpture, by Maurice Haron, was inspired by the Battle of the Curlews, referred to earlier.

Returning to the N4, it is only a short ride back towards Boyle. Rather than taking the Boyle exit to the right, proceed across the bridge over the Boyle Canal and take the path that runs along the south side of the canal for the gentle off-road journey to Boyle harbour.

County Mayo

Lough Conn
Lough Cullin
Lough Mask
Lough Carra

Mayo is Ireland's third-largest county by area, after Counties Cork and Galway. It is largely rural, with Castlebar being the county town and principal administrative centre. It is a county that invites exploration, with its varied landscape and scenery, ranging from the dramatic cliffs and white sandy beaches of its Atlantic coastline to the rugged hills and mountains, extensive boglands and large lakes of the interior.

The county name is derived from a seventh-century monastic settlement established by St Colman and a group of English monks near Claremorris at a place now known as Mayo Abbey. The Irish name for the settlement was Maigheo na Sacsan, which translates as 'the Plain of the Yew Trees of the Saxon'. The Normans conquered Mayo in the thirteenth century and installed a fledgling centralised administration, but it was not until 1570 that Mayo was formally established as a county by Sir Henry Sidney, Queen Elizabeth I's Lord Deputy in Ireland.

Mayo has a rich history. It was extensively settled in Stone Age times, and more than 10 percent of Ireland's megalithic tombs from this period are to be found in the county. It also has the oldest known field systems in the world, dating back more than 5,500 years. The story of this neolithic landscape is told in the Céide Fields Visitors Centre, located near Ballycastle in the north of the county. In the early

Christian period, and in the early years of the Middle Ages, Mayo witnessed the foundation of several influential monastic settlements and abbeys, including those at Cong, Balla, Killala, Ballintuber and Turlough, whose well-preserved ruins still stand today. The county has also seen its fair share of rebellions, wars and plantations. It played a central part in the rebellion of 1798, when a French expeditionary force landed at Killala bay and, under the leadership of General Humbert, mentioned above, took possession of a number of Mayo's main towns before being defeated at Ballinamuck, County Longford. Humbert's march through Mayo, Sligo, Leitrim and Longford is commemorated in Mayo's only official singposted cycling route, known as the Humbert Trail.

A one-time resident of Mayo provided a new word for the English language. Captain Boycott was a notorious land agent who was targetted during the Land Wars of 1879–82 in a peaceful campaign to restore land to native ownership. The campaign was organised by the Land League, one of whose leaders was Michael Davitt, who was born in 1846 in Straide, to the south of Lough Cullin.

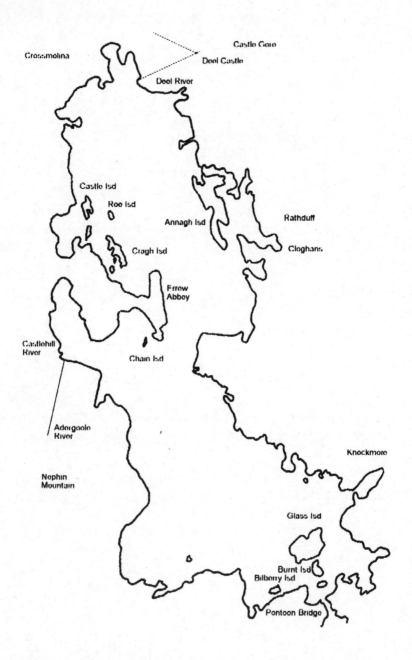

Crossmolina

Castle Gore

Deel Castle

Deel River

Castle Isd

Roo Isd

Annagh Isd

Rathduff

Cragh Isd

Cloghans

Frrow
Abbey

Castlehill
River

Chain Isd

Adergoole
River

Knockmore

Nophin
Mountain

Glass Isd

Burnt Isd

Bilberry Isd

Pontoon Bridge

Lough Conn

(64 kilometres)

LOCATION
North Mayo. The nearest principal towns are Ballina to the east and
Castlebar to the south.

LENGTH
14.5 km/9 miles

WIDTH
6.5 km/4 miles at its widest point

AREA
5,150 hectares/50 km²

PUBLIC ACCESS
Gortnorabbey pier (near Crossmolina), Errew pier, Gillaroo bay,
Pontoon, Knockmore bay, Brackwanshagh, Sandy bay, Cloghans bay

MAP
Ordnance Survey of Ireland Discovery Series Map 23, covering Mayo
(ISBN 978-1-901496-29-1), and Map 24, covering Mayo and Sligo
(ISBN 978-1-901496-28-4)

As a cycling route, the circuit around Lough Conn is of the highest quality. Despite being surrounded by mountainous terrain, the route is generally flat, with extremely quiet country roads running very close to the lakeshore for almost the entire journey. The only significant hills that you will encounter are located near the lake's southern shores at Terrybawn, but even these are not onerous. The route is dominated by Nephin Mountain (806 m) on the lake's western shores, which ensures that you do not wander off course.

While Lough Conn is one of Ireland's largest lakes, it has an intimacy and intrinsic beauty that is more commonly found in relation to smaller lakes. Yet for all its attractions, it does not appear to be well known or promoted, even among anglers. Depths in the northern, or upper, part average around 4.6 m, while the southern section is generally deeper, at up to 30 m. A drainage scheme for the River Moy in 1964 caused the winter level of the lake to fall by 1.8 m, reducing its area and linking some of what were then islands to the mainland.

Pontoon

Pontoon, which is a townland rather than a village, is a good starting point for the circuit. It occupies the narrow strip of land between Lough Conn and Lough Cullin, which are linked by Pontoon Bridge. The bridge derives its name from a pontoon that previously existed on the Castlebar–Ballina road, where the bridge now stands. In his book Irish Ghost Stories, Padraic O'Farrell relates the story of a huge black dog that is said to stand sentinel on this bridge. The dog is one of the many forms taken by the 'pooka', one of the ghostly figures of Irish legend.

The area has two hotels, the Pontoon Bridge Hotel and the more intimate and cosy Healy's Country House Hotel, both of which can be availed of for post-circuit refreshment. Setting off from the bridge in an anticlockwise direction, follow the R310 as it skirts very close to the lakeshore to Knockmore. The large island that you can see not too far off the shore is Illannaglashy, or Glass Island. The name of the island is derived from its fame as a covert location for the production of poteen, which was also stored on the island in large quantities. It was a good place to locate a poteen still, as lookouts could be relied upon to provide plenty of warning should the guards be on their way. The poteen was usually made at night, and the glow of the fires presented both a difficulty and an opportunity: the poteen makers used to light decoy fires to put the guards off the scent. In 1837, the island was described in Samuel Lewis's A Topographical Dictionary of Mayo as comprising 'good arable land with a portion of rocky pasture'. It is no longer inhabited, and all that is left on it is a number of decaying houses, the ruins of a church, and the burial place of Bishop Balefadda, who took refuge there during Penal Law times.

At Knockmore, you have the opportunity to bear left on to the quieter country roads that are an encouraging and welcome feature of most of the circuit. Pushing ahead, you pass Cloghans pier, which looks on to a sheltered bay protected by Annagh Island. A little further on is Rathduff Church, where there is an interesting Mass rock on the right-hand side of the church car park. This is known as the Carrowmore Rock and was resited to the church car park in July 1974. Mass rocks were used in Penal times and were usually located in remote areas, where a travelling priest could say Mass without coming to the attention of the authorities. The Carrowmore Rock has two distinctive bowls carved into its surface.

Castle Gore

After Rathduff, you should continue to bear left at the next junction until you come to a sharp turn right at Carragloon National School. Here you can either continue on the small lane to the left or, alternatively, follow the principal road past the school until you come to a crossroads, where you should turn left between two houses. At the rear of one of these houses are the ruins of a tall tower-like building that is completely covered in ivy. These are the ruins of a church that was built for the local landlord, Lord Arran, but the building was never used as a place of worship and fell into disrepair. Down this road, you will find the ruins of two further buildings. On the right-hand side are the ruins of an old manor house known as Castle Gore. This was built for the Earl of Arran between 1780 and 1790, after the family's original residence, nearby Deel Castle, became uninhabitable. Castle Gore was severely damaged by fire during the 1798 Rebellion. The restored building was last occupied by Colonel Cuffe, until he was encouraged to vacate the premises before it was burned down once again during the Civil War. Local legend relates that under cover of darkness Colonel Cuffe had removed all valuable furniture and artefacts from the house, prior to it being burned. He is then said to have claimed compensation from the Free State government for the burning of the house and to have included in his claim all the items that had been removed.

Deel Castle

The ruins of the older Deel Castle are in private grounds. The castle was built in the sixteenth century for the Burkes, one of the foremost Mayo families descended from the Normans. It was later granted to the Gore family, who secured the title 'Earls of Arran'. The lands surrounding the castle are much changed from the time when the Gore family was in residence. The remains of a wall that used to surround the castle, and an extensive orchard, can still be seen. The fields outside this wall are now clear pasture, but up to the early part of the twentieth century they were thickly planted with beech trees. These trees were cleared following the reallocation of the land by the Land Commission. Other remnants of the Arran estate are several stretches of narrow, ramrod-straight roads that were used by the landlord and his agents to tour the property. It is said that locals, even those who worked on the estate, were not allowed to use these roads. One local man, John Joe Kavanagh, tells the tale of how his father fell foul of this rule. His father's first job, when he was aged about fourteen, was to pump water to the upper rooms of Castle Gore using a hand pump located in the back yard of the property. On one occasion, he had been cutting turf at the opposite end of the estate from where he lived, in the townland now known as Knockfree. To shorten his journey, he ventured on to one of his lordship's roads, only to be caught by one of the stewards. He ignored the steward's entreaties to leave the road. His mother was subsequently summoned to the agent's office in Ballina and fined one shilling for her son's misbehaviour, otherwise her son would lose his job on the estate.

Several stories are told that illustrate the deep dislike for the landlord in the locality, culminating in the burning of Castle Gore in the 1920s. In one tale, a local man who worked as a labourer on the estate was peremptorily sacked for eating a turnip that he had pulled from the field on his way to work. When challenged, he told the steward that he had nothing to eat in his house and that the turnip was his breakfast. With no other job available, the man walked to Ballina, where he participated in a running race that he had the good fortune to win. The prize was a piece of material that could be made into a suit. He later sold the material for ten shillings, using the money to purchase his passage to England before moving on to Australia, never returning to his native land. Another story relates to two brothers who were expert poachers and were a thorn in the side of the landlord's agent. In a clever move, the agent persuaded them to become gamekeepers on the estate. As they knew every trick in the poaching book, they were highly effective in their new duties.

Crossmolina

Leaving Castle Deel behind, take the road that runs between Castle Gore and Deel river, and turn left to cross the bridge. This leads to another of the long, straight, narrow roads of the former Arran estate. This particular road skirts the top of Lough Conn and brings you to the N59 just east of Crossmolina. Gortnorabbey pier, on the lakeshore, lies to the south of the town, and can be accessed either off the N59 or from the R315 as you exit the town. The Deel river divides the otherwise fairly unremarkable town of Crossmolina. There are the ruins of a castle located near St Mary's Church; the castle was rendered uninhabitable in the early sixteenth century by an O'Donnell chieftain from Donegal.

Enniscoe House is located a short journey south of Crossmolina off the R315. This plain, pink-hued house, with a sweeping curved entrance road, was built around 1750 for the Jackson family and was later occupied by the Pratt family. It is now operated as a guesthouse by a Mrs Kellett. To the rear of the house, the former stables and carriage house are occupied by the Mayo North Family Heritage Centre, where expert genealogists can assist those who wish to trace their ancestry in this area.

Errew Abbey

The Errew peninsula, which stretches across the lake's middle section, is one of the more interesting and historic areas on the lakeshore. There is a fairly lengthy cycle out towards the tip of the peninsula, where you will find the preserved ruins of Errew Abbey. St Tiernan is said to have founded a monastery here in the seventh century. To access these ruins, you will have to leave your bicycle locked at the entrance gate and trek over several muddy fields for about half a kilometre. The ruins – of a building possibly constructed in around 1250 – consist of a church and cloister and a number of domestic buildings. The preservation works at the site were hindered by the fact that some of the original stones from the abbey were pillaged and used in the construction of Deel Castle and Castle Gore on the opposite shore. There is no information available as to the early occupants of this settlement, but the information panel at the site states that Errew Abbey was occupied in the early fifteenth century by a branch of the Augustinians affiliated to a larger Augustinian foundation based at Crossmolina. To the north of the site are the ruins of a small oratory called Teampallnagalliaghadoo, or the Church of the Black Nuns.

Two graves of note exist at Errew Abbey, one of a bishop of Cork and the other of an abbot of Clonmacnoise. Why they were buried at this fairly remote location, so far from the places where they practised their faith, is open to conjecture. Perhaps they hailed from this locality originally.

As you stand inspecting the ruins on this wind-blown peninsula, you may wonder why the religious chose this remote spot to establish a settlement. You have to remember that the profile of the landscape would have been very different in earlier centuries. The land surrounding the ruins would have been far more wooded than it is today, providing shelter for the abbey. In addition, the favoured mode of transport in those far-off times would have been by water, and proximity to the lake would have been of paramount importance when choosing a site. Furthermore, the Errew peninsula occupies a strategic location on this lake. There are excellent views both up and down the lake, and it is just a short boat journey to the eastern shore. A further important factor would have been the quality of the surrounding land: it is worth noting that the name 'Errew' is derived from the Irish word oireamh (not aireamh, as shown on local signposts), which translates as 'good land'.

As you leave Errew Abbey, the hill overlooking the abbey from the western side is known locally as Knockanore, which translates as 'the Hill of Shame'. Oliver Cromwell's soldiers are said to have bombarded the abbey from this hill.

Errew House

On your way down the peninsula, you cannot help but notice the large, imposing building, with a distinctive tower, sited proudly on a high point beside the road. Known as Errew House, it was built around 1870 by the local landlord, a man called Knox. He bankrupted himself while attempting to finish the building and fled to the United States. Over its life, the building has been used as a convent and a hotel; it is presently divided into self-contained apartments.

In retracing your steps back along the Errew peninsula towards the R315, watch out for a left turn that allows you to return to the main road further south, close to the village of Castlehill. At the first opportunity after you have crossed the Castlehill river, you should turn left, taking you on a course that runs parallel to the Errew peninsula. Once you pass straight through a crossroads, with the Edergoole Cemetery signposted to the left, you enter an area that is one of the true highlights of the Lough Conn circuit. (The presence of grass running along the centre of the road is an indication of how little traffic passes this way.) The area is densely wooded, and the trees meet overhead, creating a tunnel, through which you are provided with sheltered passage. It is particularly delightful during the late-autumn months, when the trees have changed colour and the road is dressed with a multihued carpet of rich autumnal colours. At the same time, you will be able to catch glimpses of the lake to your left.

Having been absorbed in and lulled by the splendour of the scene through which you have just passed, a quick reality check is provided by a couple of stiff ascents through the townland of Terrybaun. It is fortunate that these climbs are located close to the end of the circuit, as you can recover in the comfort of Healy's bar.

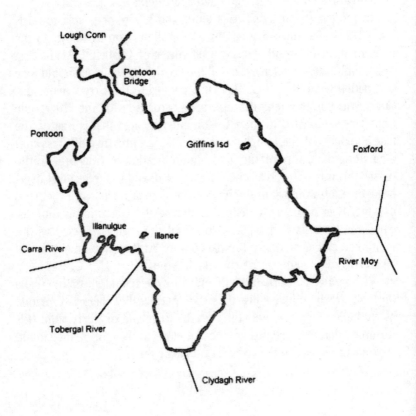

Lough Conn

Pontoon Bridge

Pontoon

Griffins Isd

Foxford

Illanulgue

Illanee

Carra River

River Moy

Tobergal River

Clydagh River

Lough Cullin

(24 kilometres)

LOCATION
North Mayo, in the west of Ireland near Foxford

LENGTH
4.3 km/2.7 miles

WIDTH
3.75 km/2.3 miles at its widest point

AREA
1,100 hectares

PUBLIC ACCESS
Access is available only from the north shore, at Drummin, where
there are several points along the Foxford–Pontoon road; at Pontoon
Bridge; and at a new mooring and access road off the
Pontoon–Castlebar Road.

MAP
Ordnance Survey of Ireland Discovery Series Map 31, covering
County Mayo (ISBN 978-1-901496-80-2)

After the experience of Lough Conn, where you can travel beside the lakeshore for most of the circuit, the route around its smaller twin is somewhat disappointing. The initial journey from Pontoon on to the Foxford Road (R318) is promising, with excellent lake views and several attractive sandy beaches along the north-eastern shore at Drummin. Unfortunately, at a fairly early stage, the road veers away from the lake towards Foxford, and you have no alternative but to follow it, so that you can cross the River Moy, which otherwise would hinder your passage further south. However, you can keep in touch with the lake and gain some relief from the traffic on the busy main road by taking the first turn right from the R318. This allows you to meander along a loop of quiet minor roads that initially run parallel to the lake, before turning sharply east and then rejoining the main road on the western outskirts of Foxford.

Foxford

Foxford is a busy town split in half by the River Moy. It is best known for its woollen mills, whose story is told in comprehensive fashion at the visitor centre located on the banks of the river. While one often hears of the divide between the two principal religions practised on the island of Ireland, the establishment and subsequent success of the Foxford mills is a lesson in practical ecumenism. While the mills were established by the Roman Catholic Sisters of Charity and driven by the entrepreneurial spirit of Mother Morrogh Bernard, technical advice was provided free of charge by a Protestant mill owner who hailed from Ulster, John Charles Smith. When the mills were founded, in 1892, the intention was to provide employment for young single women in an area where jobs were few and far between. At its peak, the Providence Woollen Mill, as it was known, employed more than two hundred people and used the fast-flowing River Moy to drive its looms. It later switched over to steam-powered looms to produce its celebrated blankets, rugs and tweed. Trading difficulties encountered during the recession years of the 1980s saw ownership of the mills change from the Sisters of Charity to a group of local businessmen. It is fitting that in its centenary year of 1992, a visitor centre, incorporating a tour of the working mill, was opened. The centre also houses a café.

Foxford is also the birthplace of the founder of the Argentinian navy, Admiral William Browne. Admiral Browne was the leader of an improvised fleet which helped end Spain's run of imperial conquests in South America. He died in 1857 and is regarded as a great hero in Argentina. In 2006, a bronze statue of Admiral Browne was brought from Argentina and erected on Sir John Rogerson's Quay in Dublin.

In order to regain contact with Lough Cullin, you will have to venture south of Foxford on the busy N58 for about 4 km, until you come to Ballylahan Bridge, where you can once again cross the River Moy. Located to the left just after the bridge are the ruins of Ballylahan Castle, sited on the banks of the Cloonlee river. This was once the home of the Norman Jordan D'Exeter. All that remains of the castle are portions of the curtain wall and one of the two rounded gate bastions. A little further on, a turn to the right begins the northbound trek back to the lake, on a quiet minor road that tracks the River Moy on its western side. After a sharp turn left, this road runs close to the southern shore, before crossing over the Clydagh river and moving away from the lake once again. There are a couple of climbs and descents before you reach a junction, where a turn right will set you on course to link up with the R310 after you have crossed, in fairly quick succession, the Tobergal and Carra rivers. The final leg of this journey brings you once again to the edge of the lake, before returning to Pontoon.

Cloon River

Srah

Partry

Ballygary
Quay

Keel
River

Partry
Mountains

Tourmakeady

Robe
River

Glensaul
River

Ballinrobe

Devinish
Isd

Cushlough

Owenbrin
River

Annagh
Point

Lusteen
More

Lusteen
Beg

Maamtrasna

Ram's
Isd

Saint's
Isd

Inishmaine

Finny

Finny
River

Ferry
Bridge

Red
Isd

Big
Isd

White
Isd

Aghalard
Castle

Upper Mask

Earl's
Isd

Ballykine
Wood

Joyces
Country

Ben Levy
Mountain

Clonbur

Cong

Lough Mask

(74 kilometres)

LOCATION
Primarily in County Mayo, with its southern tip stretching into County Galway

LENGTH
16 km/10 miles

WIDTH
6.5 km/4 miles at its widest point

AREA
8000 hectares/80 km^2

PUBLIC ACCESS
Cushlough, Cahir bay, Ballinchalla bay, Rosshill, Ferry Bridge, Annagh, Churchfield, Tourmakeady, Srah, Ballygarry and Aughinish

MAP
Ordnance Survey of Ireland Discovery Series Map 38, covering Galway and Mayo (ISBN 978-1-901496-39-0)

Lough Mask lies in a porous and cavernous limestone landscape that is geologically fascinating, and a cave diver's paradise. Sink-holes, sub-terranean rivers, bubbling springs, turloughs (seasonal lakes) and underground caves and passages combine to provide an interesting setting – and cause the lake's seasonal water levels to fluctuate greatly. Lough Mask is connected to Lough Corrib by underground streams and rivers, although attempts were made in the middle of the nine-teenth century to cut a surface-canal connection through the village of Cong. The underground rivers are visible at some locations, such as at the Pigeon Hole, 1 km north of Cong, in the centre of Pigeon Hole Wood.

General description

The lake is bounded on its well-defined western shoreline by the extensive Partry Mountains. The lands to the east are level, and the lake's eastern shoreline is jagged, with a series of promontories, bays and inlets. Upper Mask is the name given to a portion of the lake that lies in the south-west corner and stretches from east to west over a dis-tance of around 6 km. This is really a lake in itself, being almost com-pletely cut off from the main lake at Ferry Bridge. To the south of Upper Mask range the high hills of 'Joyce's Country', which stretches across to the villages of Clonbur and Cong to the east and past Maam to Killary harbour in the west. Sir William Wilde, in his magnificent book Lough Corrib: Its Shores and Islands, draws on Roderic O'Flaherty's (1629–1717) Chorographical Description of West or H-Iar Connaught, writ-ten in 1684, for the origins of the name of this district. Apparently the Joyces were a Welsh family who came to live in the region in the thir-teenth century and were under the protection of the powerful O'Flaherty clan, who ruled the western part of Connacht in medieval times – and about whom more will be said later.

The level of the lake varies by as much as 2.5 m from summer to winter. According to a survey conducted by Commissioner R. B. Beechey and Lieutenant A. G. Edye, who took soundings of the lake in 1846, the deepest part was measured at 57.3 m and is located in a lengthy channel that lies between the two islands of Lusteen More and Lusteen Beg to the west and Shintilla Rock to the east. According to the same survey, the maximum depth of Upper Mask is 24.7 m. With its varying water levels, deep channels and extensive reef formations, the lake can be a dangerous body of water from a boating perspective.

According to Sir William Wilde, the lake's Irish title provides the clue as to how it got its name. In Irish, the lake is called Loch Measca. The word measca means 'mingling' or 'in the midst of'. The waters of Lough Corrib to the south and Lough Carra to the north combine in the middle lough, so providing the name. Wilde would have been very happy with the outcome of a project to mark the millennium that resulted in the Connacht Angling Council arranging to have a stone erected on each of the main islands of Lough Mask, with the old Irish name for the island engraved on each stone.

A dark history

Lough Mask has a darker side, and its shores were the setting for a number of brutal events over the years, particularly during the latter part of the nineteenth century, when the wars against landlordism were at their height. One such incident involved Viscount Mountmorres, who, despite having been a relatively benign landlord, was murdered in Maamtrasna, on the south-western shores of the lake, on 25 September 1880. Such was the climate of fear in the region at the time that one of Viscount Mountmorres's own tenants (near whose home the viscount had been gunned down) refused to allow his body into his cottage for medical examination. Earlier, in County Donegal, the landlord of the Maamtrasna area, Lord Leitrim, was assassinated in broad daylight on 2 April 1878, despite the presence of two bodyguards. He did not have a very good reputation in the area and, according to Jarlath Waldron in his book Maamtrasna: The Murders and the Mystery, had earlier been reported as saying that he would evict twenty tenants with their families within two weeks of his arrival for the spring fishing on Lough Mask. In January 1882, Upper Mask was used to hide the bodies of Sir Arthur Guinness's elderly baliff, Joseph Huddy, and his grandson, John, who had been murdered while they were out serving eviction notices. An elderly woman had seen the bodies being dumped in the lake and had identified the location for the police. The passing on of this information was at one time connected to the subsequent savage Maamtrasna murders, about which more will be said later.

The cycling circuit around Lough Mask is majestic, and I am envious of the members of the Western Lakes Cycling Club of Ballinrobe, who regularly use the circuit for both training and events. The circuit combines lengthy flat sections with challenging climbs and descents, particularly near the village of Finny on the shore of Upper Mask. The evidence of the popularity of cycling in this region is to be found in road markings – and in the largest sign I have ever encountered on my travels warning motorists of the presence of cyclists. This is located at Srah on the north shore of the lake, just before the turn east for Partry. With the exception of a 7.5 km stretch along the N84 from Partry to the outskirts of Ballinrobe, most of the journey uses quiet secondary roads and country lanes.

Clonbur

The village of Clonbur in County Galway is an excellent base from which to set out for circuits of either Lough Mask or the much larger Lough Corrib. Previously known as Fairhill, it is located in the middle of the isthmus that is wedged between the two lakes. The recently refurbished Fairhill House Hotel, run by Eddie Lynch, retains the village's old name and provides superb accommodation at a reasonable price.

I would strongly recommend undertaking the circuit in a clockwise direction. This means that the challenging climb from Finny overlooking Lough Nafooey can be disposed of early, when energy levels are high.

Ferry Bridge

Heading west from Clonbur, the yellowy-green slopes of Benlevy rise to your left as you make your way towards Ferry Bridge, which divides Upper Mask from the main part of the lake and also marks the county boundary between Galway and Mayo. In ancient times, Benlevy used to be called Slieve Belgadain, after the Belgae people, who were one of the early groups of invaders who occupied Ireland. The Irish name for Ferry Bridge is 'Droichead a tSnámha', which translates as 'the Swimming Bridge', as the junction between the two sections of the lake used to be known as Béal a tSnámha, or 'the Swimming Mouth'. If you were to have arrived at this junction a hundred years ago, there would have been no bridge, and the only way of crossing with a horse would have been to swim.

Upper Mask

The route along the northern shore of Upper Mask provides a taste of the spectacularly rugged scenery that is a feature of the early stages of this circuit. It is hard to find words to describe adequately the unfolding scene of splendour as you progress along the shore, with the multi-hued mountains, their slopes pockmarked with outcrops of stone, completely enclosing the western end of the lake. Blackfaced sheep are precariously perched at the higher levels, while the lower slopes are striated with narrow stone-walled fields which probably yield frugal returns. Here and there, isolated cottages and farm buildings provide evidence of a small community which, even in these times, must find living tough.

Earl's Island

No islands are to be seen at the eastern end of the lake, but as you approach the village of Finny, with its distinctive whitewashed church, the road turns away from the lake and starts to rise gently, giving you a good view of the scattering of islands squeezed between the lake's narrow confines. The most westerly of these is Earl's Island. This is named after Edmond de Burgo, the son of the Earl of Ulster, who in 1338 was abducted by the de Standuns and, after a number of days spent in captivity in various locations around Lough Carra and Lough Mask, was cruelly drowned by his captors off the shore of this island.

Maamtrasna

After passing through Finny, it is easy to miss the turn right towards the pass that squeezes between the two peaks of Ben Corkey to the east and Ben Corragh to the west. This pass facilitates a route to Maamtrasna bay, another inlet off the main lake that runs parallel to Upper Mask but does not extend as far west. The road is tight and winds steeply for a lung-sapping 800 m climb, with most of your time spent out of the saddle, even on a bicycle with sophisticated gearing. The reward is an unfolding view of the meandering course of the Finny river behind you and Lough Nafooey to the west, an attractively sited lake with a not-so-pleasant Irish name. The Irish version is Loch na Fuaithe, which translates as 'the Lake of the Hatred'. This name predates the horrific events that took place on a small road to the right of the summit and catapulted the Maamtrasna area to national and, indeed, international prominence.

The Maamtrasna murders

On 17 August 1882, five members of a family spanning three generations were brutally murdered. A sixth member, a young boy, survived the attack, although he was badly injured. They lived in a primitive dwelling on a small country lane to the east of the pass. In an era of many brutal killings, the savagery of the massacre, in such a remote and impoverished location, gripped the nation and set in train a series of events that were to have repercussions for many years afterwards. The murders took place against the background of the Irish Land Wars, which saw many prominent figures in Ireland campaign against the worst aspects of landlordism, including high rents, evictions and the enforced subdivision of small plots.

Within days of the attack, ten men were arrested. The police were acting on the word of local informers – which was highly unusual in a close-knit society that normally shunned the authorities and frowned upon providing any information to them. Of the ten who were arrested, two were persuaded to turn Queen's evidence and testify against their fellow accused in return for immunity from prosecution. Following a series of trials at Dublin's Green Street, three of the accused were found guilty and sentenced to hang before the other five were persuaded to plead guilty. Initially sentenced to hang on the same day as their co-accused, they received a late reprieve for having pleaded guilty, and their sentences were commuted to life imprisonment.

The conduct of the trials, which included the suppression of vital evidence, the withholding of evidence from the defence, the intimidation of witnesses, and questions over the independence of witnesses, was subsequently held up as an example of the worse kind of miscarriage of justice. In the aftermath of the executions, it was claimed that one of the executed men, Myles Joyce, was completely innocent of the crime, as were four of the five who had been imprisoned. Influential clergymen and politicians, including Home Rule leader and Land League president Charles Stewart Parnell, called for a public enquiry. As Jarlath Waldron states in his detailed book on the subject, the ghost of Myles Joyce 'stirred up very deep waters'. A five-day debate in the

House of Commons in October 1884 ended with failure to secure an inquiry but hastened the fall of Gladstone's Liberal government the following year. A second, shorter debate, within weeks of the installation of Salisbury's Tory government, produced the same result.

Two of the five prisoners serving life sentences died in captivity. The remaining three served the full term of twenty years before being released as old and broken men, still protesting their innocence, and that of Myles Joyce.

The western shoreline

Pushing ahead, the rapid descent under the towering slopes of the Partry Mountains initially tracks east before shifting northwards to cross the Owenbrinn river. The area between the road and the lakeshore is reminiscent of one of Arthur Armstrong's highly distinctive landscapes: I wonder whether his art was inspired by this area. Should you be in need of sustenance at this stage, or just be seeking a comfortable place to enjoy the vista, Maire Luke's Scenic Bar is perched on a superior vantage point a couple of hundred metres back from the next junction.

There are many attractive and accessible shore-access points along the route to the village of Tourmakeady that are not marked on the map. Some are equipped with picnic tables and small jetties. Just south of the village is Paddy's Thatch Bar, with its appealing garden overlooking the lake. The only site of note in Tourmakeady is the waterfall on the Glensaul river to the west of the village. The waterfall can be accessed via Tourmakeady Wood (Radharc na Coille) on the southern outskirts of the village.

A scattering of islands marks the approach of the northern extremity of the lake. Views of the northern shore are limited as you make your way eastwards, crossing Derrymore Bridge, which spans the dark waters of the Cloon river. This is a favourite spot for local birdwatchers. Just before you reach Partry village and the N84, a turn to the right leads to Ballygarry quay, a popular spot amongst anglers. The initial journey along the N84 is not inspiring. There are no views of either Lough Mask or nearby Lough Carra to the east until you reach Keel Bridge, built in 1924 to span the Keel river, the link with Lough Carra. You will have to persist with the busy main road for a short while longer, but there is no need to endure it all the way to

Ballinrobe. The right of way opposite the left-hand turn towards Rocksborough leads westward through the Creagh demesne, allowing you to avoid Ballinrobe altogether. Initially, the road is rough and quite mucky in wet weather. The surface quality improves, however, after you pass an old house (which has seen better days) and take a sharp left turn. This leads to a delightful tree-lined avenue followed by a pillared entrance marking the boundary of the demesne. For those who might wish to visit Ballinrobe, the road straight ahead leads to the town centre. However, the circuit follows the road that runs sharply to the right, and you are soon crossing the River Robe and heading south. Take time to admire a magnificent refurbishment of a mill building that has been converted into a private residence, retaining the reconditioned mill wheel as a feature by the gable wall.

Cushlough bay

Cushlough bay is a well-signposted diversion that is worth visiting. The name of the bay is an anglicised version of the Irish 'Cois Loch', meaning 'Beside the Lake'. The shoreline area is the property of the Ballinrobe and Cushlough Anglers and has been tastefully developed with anglers in mind. A small clubhouse bears a plaque noting that President Mary Robinson officially opened the World Cup Trout Wet-fly Angling Championship which was held on Lough Mask in July 1994. This is in fact an annual event that has been held at this location since 1953. An information centre belonging to the Western Fisheries Region is sited to the rear of the clubhouse. The bay itself is a serene location that is well populated with swans and other birdlife. It contains the only crannóg recorded on Lough Mask, which can be seen to the right of the bay.

The First Battle of Magh Tuireadh (Moytura)

Tracking back from Cushlough bay and resuming a southerly path brings you through an area that, according to Sir William Wilde, was the scene of the final stages of the First Battle of Magh Tuireadh (or, to use the modern version, Moytura), during which the Tuatha Dé Danann defeated the Firbolgs. Mention of this battle, and the debate over its location, has already been made in the description of the circuit around Lough Arrow in County Sligo (see page 167). While most authoritative sources now discount Wilde's theories in relation to the location, his book nevertheless provides a lengthy description of the battle and its defining moments, and links these to some of the many monuments that populate this area, particularly in the triangular sector bounded by the villages of Cong, Neale and Cross. It is not proposed to cover these within this book, as they are a fair bit off our route, but for those who are interested I can recommend Wilde's entertaining book. However, you cannot miss the great grassy mound that rises to the left of the road south of the turn for Inishmaine Abbey. This cairn commemorates the spot where King Eochy (sometimes written as Eochaidh) Mac Erc is said to have perished. Eochy, the last of the Firbolg Kings of Erin, died after slaying three sons of one of the leaders of the opposing Dé Danann forces. He is not to be confused with Eochaidh Mac Maireadha, after whom Lough Neagh is named. If you wish to climb the mound, from which there is a great view of the surrounding area, there is a path to the left of the road leading to the cairn, or alternatively head towards Killour and, after a further kilometre, take the first turn left. Another kilometre further on, there is a turn right that leads to the base of the cairn.

Inishmaine Abbey

An interesting and historical peninsula that was once an island lies west of the road that runs past Eochy's Cairn. The road down the peninsula is known as Boher-na-Corp, or 'the Road of the Corpses', and leads to Inishmaine Abbey. Access to the ruined abbey is via a track that runs through private farmland, so be sure to shut the gate at the entrance to the track. As you approach the abbey, the rough track gives way to a far more solid base of smooth limestone slabs that slope towards the water. If you are travelling along this path in the wetter winter months, caution should be exercised, as it is very easy for your wheels to give way on the sloping surface, causing a nasty tumble.

Wilde described the abbey in 1867 as 'a charming group of ruins, standing by the water's edge, surrounded by well-grown timber' and as 'one of the most beautiful churches in Ireland'. The surrounding 'timber' has largely disappeared, and the ruins have a rather forlorn appearance, with the grounds of the surrounding enclosure now regularly trodden upon by livestock, making access quite mucky in winter. The ruins appear to have deteriorated since Wilde's time – and very much so since 1779, when a drawing by Angelo Maria Bigari (which is included in Dr Peter Harbison's Guide to National and Historic Monuments of Ireland) depicted it in a considerably more substantial state than that in which it exists today. The walls are no longer ivy-clad, as portrayed in both Bigari's drawing and Wilde's later illustration, but there is still much to admire among the ruins. The carved pillars which would have supported the chancel arch are still there, and the carved figures at the extremities of the mouldings are still visible, though they are not as clear as they were in Wilde's day. The church is fairly small, measuring only 19 m in length, and entrance is by a door in the west gable wall. Fortunately, a cattle grid has been installed at the entrance to ensure that the livestock do not damage the interior. It is a pity that a similar grid is not placed at the pedestrian entrance at the north end of the enclosure. Along the western end of the enclosure wall is a gatehouse, which has been partially restored. You can climb a small staircase to a platform in this building, where you get a slightly elevated view of the surrounding area.

I am pleased to report that there is an information panel at the site providing a limited history of the abbey. It is thought to have been built on the site of a small church that St Cormac built in either the sixth or seventh century. According to Wilde, the church was sited on the wasted foundations of a stone fort that had been occupied by King Eoghan Beil, king of Connacht. The abbey dates back to the early thirteenth century and, according to the information panel at the site, it was a Benedictine Monastery, although Dr Peter Harbison has attributed it to the Augustinians. He also reports that it was burned in 1227 by Hugh, son of Roderic O'Connor. It is likely to have been used after that date, and the information panel indicates that at one stage an order of nuns occupied the site.

The shoreline near the abbey is very rugged, with large pock-marked limestone slabs sloping towards the lake. A similar pattern is evident along the length of this peninsula, the end of which can be reached via a right of way through privately owned farmland. Once again, the importance of closing gates encountered along the track must be stressed.

The Cong Canal

Returning slightly inland to the main route of the circuit, the journey south brings you to Carrownagower Bridge, which spans the Cong

Canal. This 6 km-long waterway was originally conceived as a combination of a navigational link between Lough Mask and Lough Corrib and a winter-flooding-relief facility for Lough Mask. It was started in the mid-1840s as a famine-relief project and was beset by a host of problems, from structural to financial. Ironically, one of the problems was a shortage of labour as high emigration took its toll on the local workforce. The fact that the waterway was cut through a porous limestone landscape resulted in great difficulty and expense being involved in sealing the channel cut. As the canal was nearing completion after five years of digging, its viability was called into question. The expansion of railways into the west of Ireland and a decline in the appeal of waterborne transport combined to reduce the viability of the project, and it was subsequently abandoned without being fully tested. While there is a healthy flow of water under Carrownagower Bridge and all the way west to Lough Mask, particularly in the winter months, the canal bed from Drumsheel on the outskirts of Cong village is dry and overgrown with mature trees and bushes.

Aghalahard Castle

As you continue south towards Clonbur, you have the opportunity to take a slight diversion to the left, which will link up with the Clonbur–Cong road a little further east than the main junction. This diversion affords the opportunity of a brief visit to Aghalahard Castle, a three-storey fifteenth-century tower house standing on the county boundary between Mayo and Galway. Wilde describes the structure as 'exceedingly well built' but this has not stopped the ravages of time and perennial neglect from taking its toll since his day. One half of the tower has fallen away, but you can still appreciate some of its architectural features. It is not too dissimilar in style and features to the well-preserved Aughnanure Castle on the western shores of Lough Corrib. The castle was owned by the MacDonnell family until it was sold to Sir Benjamin Guinness in the nineteenth century.

The Pigeon Hole

As you emerge on to the Clonbur–Cong road, there is another small detour which can be undertaken before returning to Clonbur. The purpose of this detour is to visit one of the unusual natural phenomena that abound in this area – and a place which I mentioned earlier as being a site where you can view one of the underground rivers that link Lough Mask and Lough Corrib. This is the Pigeon Hole, and it is to be found in a wood of the same name located a short distance to the south of the main road, slightly east of where you rejoined that road. It is fortunate that it is now surrounded by fencing, as it would be quite easy to stumble into this dark, cylinder-shaped crevasse, cut deep into the limestone rock. A series of sixty-one steps leads down into the bowels of a damp cavern, where pools of clear water are fed by underground streams. A torch would be useful when exploring the interior.

It is a pity that three unattractive pipes had to be installed to draw water from the hole. It is hard not to be startled by the sudden unheralded movement of these pipes when the pump at the surface is turned on. Drawing water from the hole is in keeping with a long tradition, although in pre-mechanised times it was a far more sedate affair.

One of the first questions most people ask about this site is why it is called the Pigeon Hole. Apparently the name is derived from the large number of pigeons that were seen flocking in and around the hole in bygone days.

Ballykine Wood

The area to the right of the road to Clonbur is Ballykine Wood. There are some beautiful off-road trails that lead through this wood to the lakeshore and link up with another wooded area, Rosshill, that lies to the north of Clonbur. These trails are in fact connected to the woods that surround Ashford Castle, which is covered in the section dealing with Lough Corrib (see page 260). It is possible to reach Ballykine from Ashford Castle without having to venture on roads used by cars, and the trails are therefore ideal for walking enthusiasts. Cyclists are also welcome in both Ballykine Wood and the aforementioned Pigeon Hole Wood.

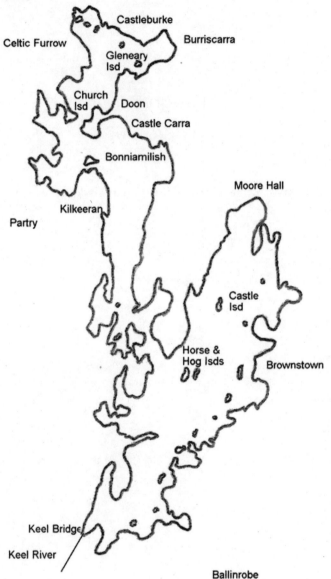

Ballintubber
Abbey

Castleburke

Celtic Furrow

Burriscarra

Gleneary
Isd

Church
Isd

Doon

Castle Carra

Bonniamilish

Moore Hall

Kilkeeran

Partry

Castle
Isd

Horse &
Hog Isds

Brownstown

Keel Bridge

Keel River

Ballinrobe

Lough Carra

(47 kilometres)

LOCATION
County Mayo, to the north-east of Lough Mask

LENGTH
9.7 km/6 miles

WIDTH
1.6 km/1 mile at its widest point

AREA
1,600 hectares/16 km^2

PUBLIC ACCESS
The only official public access is at Kiltoom near Moore Hall on the eastern shore, although there are a number of informal shore-access points.

MAP
Ordnance Survey of Ireland Discovery Series Map 38, covering Galway and Mayo (ISBN 978-1-901496-39-0)

Lough Carra comprises two separate bodies of water divided by a sub-stantial peninsula, with a narrow channel linking the two sections. It is often overshadowed by the larger Lough Mask, to which it is linked by the Keel river. Indeed, its southern section is only divided from the northernmost reaches of Lough Mask by a relatively narrow strip of land. It is the most northerly of the chain of three lakes that at one time facilitated waterborne travel from the sea at Galway deep into the heart of County Mayo. In centuries past, its surface area was more extensive: in the thirteenth century, it is said to have extended right up to the walls of Ballintubber Abbey, almost 2 km to the north of its present northern boundary. It is not a very deep lake, although depths in Black Hole bay near the Doon peninsula reach 18 m in places.

The lake's most striking feature is its strange colour: it is a pale pel-lucid green over a white marl bottom, and makes quite a contrast to the usual colour of the water in our lakes, which ranges from a som-bre black or peaty brown to a whitish blue. Lough Carra is mainly fed by lime-rich underwater springs, whose bubbling waters carry dis-solved limestone, creating a chalky marl-clay residue on the bed of the lake. This sediment is not conducive to the growth of the aquatic plants that one normally sees in the waters of Irish lakes, so the water clarity is extremely good.

There is some debate as to the origins of the lake's name. Some have argued that it is derived from Cera, who was the daughter-in-law of Nemhedh, an early foreign invader. A more likely source is the old Irish name for the lake, 'Foinloch Ceara', which translates as 'Drifting Snow', referring to its white lake bed.

The circuit around Lough Carra is a very mixed bag. The western shore is somewhat disappointing because the route uses the busy N84. This is the main road between Galway and Castlebar and its surface is quite bumpy, despite its national-road status. Views of the lake from this road are few and far between but there is a limited loop through the townland of Kilkeeran, skirting the village of Partry, that brings you closer to the lake for a short while. Fortunately, only about one-third of the circuit is undertaken on this road. For the remainder, the circuit avails of very quiet secondary roads that afford frequent excel-lent views of the lake, particularly around Burriscarra and Muckloon. The northern part of the lake is also its most historic and interesting area, with plenty for the visitor to investigate.

Moore Hall

A good starting point for the lake circuit is the car park at the front of Moore Hall on the eastern shore. The lands around Moore Hall are now owned by Coillte. There are some beautiful trails through the forest that encircles the ruins of the house and across the road through Kiltoom; you can inspect these ruins either at the start or the end of your journey. It should be noted that cycling is permitted in Moore Hall Wood.

Moore Hall was the family home of the Moore family, whose members made a significant contribution to local and national politics and to Anglo-Irish literature down through the years. Their house is now in ruins and stands eerily vacant in the midst of an encroaching forest. It was built by George Moore (1729–99), who claimed descent from Sir Thomas More, Henry VIII's chancellor, who was beheaded on the king's order in 1535. George Moore had made a fortune in Spain and returned to Ireland in 1790 to build a house on land he had purchased on the shores of Lough Carra seventeen years earlier. The house was completed in 1796.

George Moore's son John (1767–99) was appointed President of the Provisional Republic of Connaught during the French invasion of 1798, referred to earlier. He accompanied General Humbert on his revolutionary march through the west of Ireland but was captured at Ballinamuck, County Longford, when the rebels were defeated. He died in captivity in Waterford one month after the death of his father. His remains were reinterred at the Mall, Castlebar, with full military honours in August 1961.

In contrast to some of their more notorious landowning counterparts, the Moores appear to have been benevolent and compassionate landlords. In the middle of the nineteenth century, the family developed an interest in racehorses – an interest which was to bring both tragic and beneficial results. Augustus Moore (1817–45), a grandson of the builder of Moore Hall, was killed at the age of twenty-eight when he fell from a horse called 'Mickey Free' while competing in the 1845 Grand National. His older brother, George Henry Moore (1810–70), decided that the family's interest in horses could be used to relieve the plight of his stricken tenants during the Great Famine of the 1840s. In 1846, he entered his horse 'Coranna' for the Chester Cup in England and placed some large bets on him to win. He is reputed to have told the jockey, Frank Butler, that he was competing not only against the other horses in the race but also against the 'Four Horsemen of the Apocalypse'. When the horse won, George cleaned up, making £17,000 in prize money and winnings, a massive sum in those days. He used this money to charter a boat to import four thousand tonnes of maize, which was distributed to his needy tenants. It is said that no tenant of the Moore estate died from hunger during the Famine period. He later became a Member of Parliament and was active in campaigning for tenants' rights.

Perhaps the most famous of all the Moores was George Augustus Moore (1852–1933), who, despite not having received a formal education, became a prolific writer and was a key participant in the Irish literary revival of the early part of the twentieth century. He spent his formative years in Paris, where he rubbed shoulders with some of the leading impressionist painters, including Monet, Renoir, Degas and Manet, as well as as the novelist Emile Zola, who was an influence on Moore's early works. Moore returned to Ireland in 1901, taking up residence in Dublin, where he associated with all the leading lights of the city's artistic and literary communities, including W. B. Yeats, George Russell, Walter Osborne, Douglas Hyde and Lady Gregory. He later moved to London, where he died in 1933. His cremated remains were interred on Castle Island, which can be seen from the shore at Kiltoom.

It is strange that the ancestral residence of a family that was well thought of in its community and which had provided relief to a stricken populace during the Famine should be treated in the same way as the great houses of those who had been less generous to their tenants. Moore Hall suffered the indignity of being burned by anti-Treaty republican forces in 1923, and the shell of this fine Georgian building remains forlorn and neglected to this day. To the rear of the house there is an unusual tunnel yielding a rear access.

Burriscarra Friary

Leaving Moore Hall car park, turn right to embark on an anti-clockwise circuit. There is an early opportunity for a small extension by continuing in a southerly direction at the first junction and looping around the peninsula that extends to Otter point. Returning to the junction, the road north accesses an area that is both extremely interesting and profoundly historic. The first antiquity of note that you encounter is Burriscarra Church and Friary. The friary was built at the end of the thirteenth century by Adam de Standun for the Carmelite Order but was abandoned after less than a century, in 1383. De Standun was an Anglo-Norman who hailed originally from Warwickshire in England. It was taken over and refurbished by the Augustinians in 1413 but was damaged by fire in 1430, after which it was repaired and extended. The ruins are well presented and in good order. There are several graves within the main part of the abbey, following a practice that is common with churches and abbeys that have been deconsecrated. The ruins of a parish church lie nearby, surrounded by a graveyard that is still in use today. These ruins date back to the fifteenth century, when the Augustinians acquired the abbey.

Castle Carra

If you continue down the lane that runs past Burriscarra Friary, you can reach the castle that was named after the lake. It too was built by Adam de Standun towards the end of the thirteenth century and occupies a dominant position on a prominent eminence on the lakeshore. Access to the castle can be difficult, particularly during the winter months, as the lane gets extremely muddy and the fields and wooded area surrounding the castle are treacherous underfoot. It is worth persisting, however, as the ruins of the castle are in a reasonable state of preservation. Iron gates prevent access to the building itself, but you can wander around the rampart walls and enjoy the magnificent views that its inhabitants had down the main body of the lake and across to the Partry Mountains, which dominate the western landscape. While this location now appears remote, you quickly appreciate why de Standun chose this spot for his castle: the site occupies a highly strategic position, and there would have been plenty of warning of approaching waterborne enemy forces. Today it is more exposed than it would have been in his time, when the waters of the lake would probably have lapped against the stone ramparts on its southern side. It would also have been surrounded by fairly dense woodland, offering protection on the shore side.

The lake and shoreline around Castle Carra are interesting. If you take a position on the castle ramparts and look westward towards Doon peninsula, you will be able to discern the darker shadows on the surface of the water that indicate deeper water. This is the deepest part of the lake and is aptly named 'Black Hole'. To the left of the castle is a small, rectangular bay that must have been useful for docking boats. In more recent years, its gently shelving bed and appealing clarity made it a popular venue for swimming, and if you search closely it is possible to see the base of a diving board, which no doubt provided many hours of athletic pleasure for its users.

Over the centuries, the castle passed through various hands before being granted to Sir Henry Lynch, Baronet of Ireland, under the Acts of Settlement of 1678. Unlike the Moores, the Lynches were not known as benevolent landlords. By way of illustration, local folklore tells of a family who were tenants of the Lynches, living near the village of Partry on the western shore of the lake. It is said that they were evicted on Chrismas Eve for killing a hare.

Doon peninsula

Retracing your steps along the lane from Castle Carra, you pass by the access point for a remarkable and spectacular area that can be easily missed, Doon peninsula. I was fortunate during one visit to meet up with Tom Quinn, who owns the 10.5 hectares of peninsular land and who has been developing what he describes as 'the Doon experience'. The area is a site of national importance and contains evidence of lakeshore habitation extending back to the Bronze and Iron Ages. The peninsula has a pivotal strategic location at the narrowest point along the lake. Church Island, on the western shore, lies very close, and any waterborne traffic would have had to negotiate a very narrow channel between these promontories. It has been established that Doon was the location of a promontory fort, with as many as thirty habitation sites scattered around the peninsula. Late Bronze Age ramparts have been uncovered: these would have provided the inhabitants with formidable defensive cover at a time when tribal conflict and invasion were the norm. Several stone projections and pillar-type stone structures have also been discovered, and there have been various interpretations as to their use or purpose.

Doon also possesses an interesting cave that found use as a refuge for a priest in Penal times (and, in more recent times, for IRA activists on the run from the Free State army during the War of Independence). The cave is known as the 'Priest's Hiding Hole'.

During the latter half of the seventeenth century and the early days of the eighteenth century, the 'Penal Times', Catholics were not allowed to own land, their bishops were exiled, and their priests were required to register in court. In some places, Mass was said in remote places, at great personal risk to the priests. One such place was Church Island, the 3.2 hectare island across the narrow channel from Doon. There was a notorious priest-hunter in the area called Seán na Sagart (sagart being the Irish word for 'priest'), whose real name was John Mullowney. When Mass was being said on Church Island, some of the congregation were detailed as lookouts, to ensure that Seán na Sagart did not creep up unannounced. If he approached, the priest was quickly whisked off the island by boat and secreted in the cave on Doon peninsula. Ironically, Seán na Sagart is buried in the cemetery adjoining Ballintubber Abbey – which you will visit later – where he has several members of the clergy for company.

Almost due south from the tip of Doon peninsula is a small island with a delightful name, Bonniamillish. This is an anglicised combination of two Irish words, báinne, meaning 'milk', and milis, meaning 'sweet'. It is believed that this name was given to the island because of the sweetness of the milk supplied by the cows that grazed there and is attributable to the consumption of a herb that grows on the island.

Church Island

Church Island, which occupies a strategic location across the 12 m-deep narrow channel from Doon, has an interesting history. Recent tests have revealed that there was a human presence on the island as far back as 3000 BC. The island has had many names over the years, including Finan's Island, Cummin's Island and Shrine Island. St Finan is believed to have established a church on the island in the sixth century, and the skeletal remains unearthed during excavation of a small portion of the island may well be his. A small fourteenth-century church has recently been restored, and basic accommodation, in the form of small chalets, has been installed. The island is now used a centre for reflection and pilgrimage, with visits organised from Ballintubber Abbey.

Castleburke

There is a fine view from the northern shore of Doon of another of Lough Carra's Anglo-Norman castles, Castleburke. This can be reached by resuming the road circuit for a short ride north from Burriscarra and taking the first turn left, with the towering summit of Croagh Patrick looming directly in your line of sight ahead. The castle was previously known as Killboynell Castle and is thought to have been built by the O'Flahertys. It was acquired by the Burkes of Mayo in the late sixteenth century. The Burkes of Mayo were descendants of a powerful Norman called William Fitzadelm De Burgo, who arrived in Ireland with Strongbow, Earl of Pembroke, in 1172. They are sometimes referred to as the Lower Mac William Burkes, to distinguish them from the Clanricarde Burkes, whom you will have encountered during your travels around Lough Derg as the people who built Portumna Castle.

The man who acquired Castleburke was Theobald Burke, better known by his Irish name of Tiobhóid na Long, which translates as 'Theobald of the Ships' – because he was a noted seaman and owned a good many ships. In that respect, he was following in the footsteps of his mother, Grace O'Malley, better known as Gráinne Uaile, a pirate queen much celebrated in Irish folklore and song. Theobald was knighted for coming to the assistance of the English at the Battle of Kinsale in 1601. In 1627, Charles I elevated him to the hereditary title of First Viscount Mayo. He did not have long to enjoy this title, as he was murdered two years later by his brother-in-law, Diarmeen O'Connor, while on his way from his castle to nearby Ballintubber Abbey. He is interred in a tomb in the sacristy of the abbey.

The Burkes continued to reside in the castle for many years. Though now in ruins, it is still owned by a member of the Burke family and has one of the best views of the lake from its northern shores. The buildings to the side of the castle are known as the Hangman's House.

Ballintubber Abbey

Stone walls dividing small fields dominate the landscape as you make the short journey from Castleburke to an area of religious significance dominated by a church that is known as 'the abbey that refused to die'. Ballintubber Abbey has survived two fires, suppression, attack and partial destruction, but in the midst of its changing fortunes down through the years it has retained a unique position among Irish religious establishments. It is the only church in Ireland founded by an Irish king that is still in daily use: from the time of its foundation, Mass has been celebrated within its walls without a break. It should be noted that Tuamgraney Church (on the shores of Lough Derg), which predates Ballintubber Abbey by over two hundred years, lays claim to be the church in longest continuous use, but in my view Ballintubber's claim is stronger because Tuamgraney today houses a heritage centre, and services are not held there on a daily basis.

On one visit to Ballintubber Abbey, I had the good fortune to meet the parish priest, Father Frank Fahey, an engaging and energetic man, and a great storyteller with a tremendous fund of local knowledge that he loves to share. Over many years, he has dedicated himself to the continued restoration of the abbey and the enhancement of the 'Ballintubber experience' for visitors. The abbey has been very fortunate in having been under the stewardship of Father Fahey and his predecessor, Father Thomas Egan, who between them have presided over its restoration with great care and dedication over the last forty years.

Ballintubber is an anglicised version of the Irish 'Baile an Tobar', which translates as 'the Town of the Well' and refers to a well that is believed to have been used by St Patrick in 441 to baptise local converts to Christianity. In that year, on his return from Croagh Patrick, St Patrick built a small church on the site (which now houses the abbey) where he is said to have prayed for forty days and forty nights. St Patrick's well is now to be found in the midst of the par-three golf course adjacent to the abbey grounds.

The abbey was founded in 1216 by Cathal Crobhdearg O'Conor, king of Connacht, also known as 'Cathal Mór of the Wine-red Hand'. Local folklore suggests that it was built as a reward for kindness shown by a local man called Sheridan to Cathal before he became king. One of the legends associated with the abbey states that the king's builders first built it in the wrong place. There is another Ballintubber in County Roscommon, associated with St Brigid, and apparently the builders mistakenly constructed the promised church there instead of in County Mayo. When the king learned of the mistake, he undertook to erect a much grander church on the correct site, and it is this which still stands today.

The abbey was initially dedicated to the Holy Trinity and was occupied by the Canon Regulars of St Augustine. The Canons were secular priests who first came to Ireland early in the twelfth century and lived in community following rules laid down by St Augustine. The Annals of Loch Cé record that the first abbot was Maelbhrighde O'Maicin, who died in 1225. Despite a fire in 1265, which destroyed the nave, the abbey prospered through the following centuries, enlisting into its community men of noble birth who brought with them vast tracts of land. However, the Act of Suppression of 1534, and King Henry VIII's assumption of the position of head of the Church in Ireland in 1537, led to Ballintubber Abbey being surrendered to the king in 1542. Like some of the more remote monasteries in the west of Ireland, such as Boyle Abbey in Roscommon, Ballintubber continued to operate late into the sixteenth century. In 1603, the abbey lands were granted to John King, the same man who had succeeded to the lands of the MacDermots, Lords of Moylurg (see page 199).

The abbey was taken over by the Augustinian Friars in 1635, but in 1653 Cromwellian soldiers attacked it and destroyed many of the monastery buildings, including the cloisters and domestic quarters. Remarkably, the abbey church survived, although its timber roof was burned. Mass continued to be said in the church for more than two centuries, including through the Penal times, even though the structure was roofless. In 1846, the Archbishop of Tuam appealed for the abbey to be restored, but it was a difficult time economically for the country, with famine preoccupying the populace. Restoration finally got under way in 1889, concentrating on the church. In more recent times, this has continued, with the Chapter House restoration being completed in 1994.

The abbey celebrated its 750th anniversary in 1966, and the event was marked by the issuing of two stamps depicting an 1845 engraving of the church. The simplicity and austerity of the abbey's bright whitewashed interior may surprise some visitors.

Pilgrim's Way and Croagh Patrick

From Ballintubber Abbey, there is an ancient roadway that runs for 35 km to Croagh Patrick. The roadway follows the path of a chariot road that was a branch of one of the main ancient roadways, known as slighe, that facilitated travel by land more than two thousand years ago.

The roadway from Ballintubber is known as 'Tóchar Phádraig' and is still used today by pilgrims, who retrace the steps of the many people who have visited Croagh Patrick. In the abbey grounds are the ruins of a hostel where pilgrims used to wash their feet on their return from Croagh Patrick. Unfortunately, no such amenity exists for the modern pilgrim.

The Celtic Furrow

Resuming the circuit, turn left after leaving the abbey on to the road that links up with the N84. This road coincides with the early stages of the Pilgrim's Way. At the junction with the N84, there is a white-washed cottage, alongside which is the Celtic Furrow Visitor Centre. The centre, the brainchild of Father Fahey and a number of his parishioners, was opened in 2002. The centre features an extensive exhibition that traces Ireland's heritage, both physical and cultural, back to neolithic times. It links pagan and Christian traditions and customs, and shows how they are still an important influence today. The exhibition even attempts to carry this theme forward and looks at the choices faced by the increasingly materialistic culture in which we live today. The exhibition is both informative and thought-provoking and is well worth a visit.

Pushing ahead, the journey down the western side of Lough Carra cannot compete with that along the eastern shoreline, as coping with the busy main road absorbs a good deal of your attention. Just before you reach the village of Partry, there is a turn to the left that facilitates a brief escape from the main road, enabling you to get closer to the lakeshore through the townland of Kilkeeran. The road emerges south of Partry and you will have to remain on the N84 until you have crossed Keel Bridge, which spans the river that links Lough Carra to Lough Mask. About 1 km south of the bridge, turn left where it is signposted for Rocksborough, and once again you will find yourself on quiet, winding roads that are pleasant for cycling. You will have to pay close attention to the map in order to ensure that you do not find yourself meandering down a lane that leads to a dead end. Views of the lake are sporadic, but if you follow the signposts for Brownstown, you can embark on a loop that goes very close to the shore. After Brownstown, watch out for a sharp left turn that accesses the road that runs parallel to the lake and leads you back to Moore Hall.

County Galway

Lough Corrib

Galway is Ireland's second-largest county by area and incorporates Ireland's second-biggest freshwater body of water, Lough Corrib. The lake separates the county's more fertile eastern plains from its rugged and more popular western region. The region of Connemara to the west of the county boasts some of the finest scenery to be found on the island of Ireland.

The county's name is derived from the Irish word 'Gaillimh', which translates as 'Stony River' and refers to the vigorous River Corrib, which flows into the sea at Galway bay. Galway city, the county's administrative capital, is the fourth-largest city in the Republic of Ireland and is known as 'the City of the Tribes'. This refers to the city's celebrated medieval history as a thriving and prosperous port, with strong trading links to Continental Europe, and Spain in particular. Fourteen local merchant families, or 'tribes', were said to have controlled the town and its port, hence the name. One of the city's best-known landmarks, the Spanish Arch, dates from this period.

Connemara is blessed with an abundance of unspoiled country-side dominated by rugged mountains, where cascading streams constantly replenish the plentiful rivers and lakes. Quiet roads meander through a rough landscape, where wet bogland abounds and stone-walled fields enclose poor-quality land that requires back-breaking labour to yield any return. Often, the grazing of black-faced sheep is all that the land will support. According to Sir William Wilde, the name 'Connemara' is a derivative of the term 'Conmaicne-Mara', used to describe territory belonging to Conmac, a descendant of Queen Maeve of Connacht. This part of the county is host to Ireland's largest Gaeltacht region, where the Irish language is still in daily use and where Irish heritage, culture and folklore are carefully preserved.

   To the east of Lough Corrib, the landscape is less scenic but more bountiful. The county's territories stretch far inland, right across to the River Shannon, and some way south, along the shoreline of Lough Derg. The modern-day tourist will usually pass quickly through east Galway as they gravitate towards Galway city, Connemara and the Aran Islands. However, the eastern region hosts a number of thriving market towns, such as Tuam and Ballinasloe.

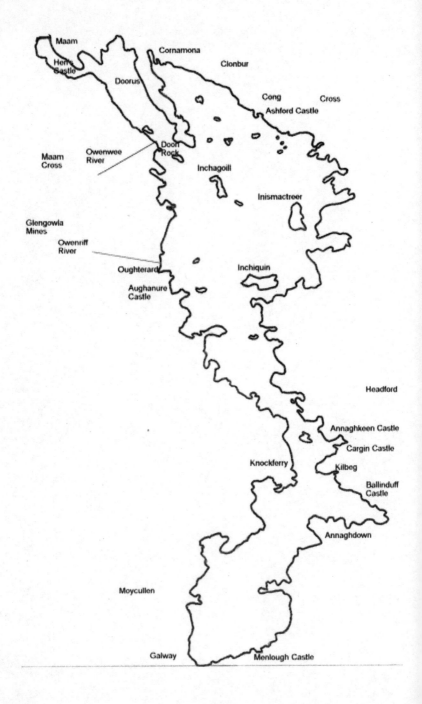

Lough Corrib

(200 kilometres)

LOCATION
County Galway, stretching from Galway city to Maam Bridge

LENGTH
43 km/27 miles

WIDTH
Varies greatly, but 16 km/10 miles at its widest point

AREA
17,600 hectares/17.6 km²

PUBLIC ACCESS
Upper Corrib – There are numerous car parks, public quays and slip-ways, including Derrymoyle, Oughterard pier, Hackett's pier, Birchall, Collinamuck, Knockferry, Rinnaknock, Greenfields, Derries, Golden bay, Lisloughrey and Cornamona. Lower Corrib – Annaghdown pier, Streamer's quay and Galway city.

MAP
Ordnance Survey of Ireland Discovery Series Map 38, covering Galway and Mayo (ISBN 978-1-901496-39-0), and Map 45, covering County Galway (ISBN 978-1-903974-10-0)

Ireland's Ice Age was generous in the legacy it left in the west of Ireland. However, of all its gifts it would be hard to surpass Lough Corrib for sheer natural beauty. Often beauty and size do not combine well, but in the case of this lake the marriage is sublime. Its scenery ranges from the wild and dramatic mountainous landscape that borders its northern and western perimeters to the more sedate shoreline adjoining Galway's flat eastern plains. Lough Corrib is Ireland's second-largest lake in terms of surface area and its shoreline is over 200 km long. It is studded throughout with many islands, some of which are quite large and inhabited, while others are small and flat and just about break above the surface.

Profile

The lake naturally divides into two parts: the lower lough, to the south, which stretches in a northerly direction from the outskirts of Galway city to Annaghdown, and the upper lough, which slants off in a north-westerly direction from that point, extending almost as far as Maam Bridge. Further subdivision is necessary to describe in more detail the lake's most striking natural features. The extreme north-western area, where glacial drift gouged out a narrow and deep basin with high hills on either side, is the most dramatic. From the tip of Doorus peninsula, the waters of the upper lough expand to take on the shape of an inverted triangle, with the widening body of water broken up by numerous islands. To the west and north, hills continue to dominate the landscape, but in the east the shoreline merges with the generally low-lying plateaus of the eastern Galway plains. The two areas of the upper lough are the deepest parts, with pockets descending as far as 46 m. The middle part of the lake is a long, narrow channel studded with islands and bound by a jagged shoreline, with many small bays on both shores. Another expansion leads to the reed-fringed southern basin. This is the shallowest part of the lake, with an average depth of only 2 m to 3 m.

The lake has an abundant water supply. Its principal source is from Lough Mask through a series of underground streams that course through the cavernous limestone isthmus that lies between the two lakes. This is augmented by the vigorous flow of the many rivers and streams that drain the rain-drenched slopes of the Connemara mountains to the west and the 'Joyce's Country' mountains to the north. A significant supply is also provided by the Shrule and Clare rivers to the east. The lake empties into the Atlantic Ocean at Galway bay via the River Corrib and also supplies much of the drinking water for the region.

Legendary origins

According to Sir William Wilde, the name 'Lough Corrib' has a legendary source. He states that its old Irish name was 'Lough Orbsen', after Orbsen Mac Alloid, one of the many names used by Manannán Mac Lir when he assumed mortal form. Manannán was one of the most popular deities in Irish mythology, and was lord of the sea – under which could be found two other important mythological territories, the Land of Youth and the Islands of the Dead. He was the guide used to access these lands. Lough Corrib's vast size, and its sometimes turbulent waters, made it appear to be an inland sea, and therefore an appropriate expanse of water over which Manannán could preside. The word 'Corrib' is apparently a corruption of 'Orib', a variation of the name 'Orbsen'. According to legend, Manannán's mortal form engaged in a battle with a warrior called 'Uillen Red Edge' at Moycullen on the western shore of the lake. Unfortunately, he was slain, and it is said that the waters of Lough Corrib sprang up at the spot where his blood was spilled.

Early shoreline occupation

From the earliest times, Lough Corrib represented a formidable frontier in western Ireland. While no crannógs have been identified on the lake itself, the remnants of many ringforts can be found close to its eastern shoreline. The pattern was not repeated west of the lake. Apart from the natural barrier presented by the lake itself, the hard quartzite of the hills on the northern and western shores would have precluded the raising of these structures in those areas. After the Normans finally made their way across the River Shannon, one of their leaders, De Burgo, forced the powerful O'Flaherty clan to retreat across the lake, having ceded their lands. Up until then, the O'Flahertys had been vying with the kings of Connacht, the O'Connors (also written as 'O'Conor'), for supremacy in the west. However, they subsequently established a powerful and dominant base in west Connacht and were greatly feared in the port of Galway to the south. Further north, the

Joyce family dominated the territory around the north-west of Lough Corrib that still bears their name. In the thirteenth century, castles began to appear on the eastern shore of the lake, particularly along the narrow middle section, as the Normans defended their recently acquired lands from waterborne raids across the lake. Later, quite a few tower houses followed, but both these and the older castles were wrecked during the Elizabethan and Cromwellian wars. Several ecclesiastical sites, such as those at Annaghdown and Ross, suffered the same fate. Some of the tower houses have been restored, but the ruins of many others provide plenty of interest to the modern-day lake explorer.

Navigation

Despite the dangers of many concealed reefs and shoals, Lough Corrib was used for many years for the transport of both people and goods. Man-made canals played a role in improving the lake's accessibility and connections. The earliest canal in Ireland is thought to be the Friar's Cut, which is said to have been built prior to 1150, so that the friars of Claregalway Abbey would have an easier passage to Galway via Lough Corrib. In 1848, construction began on a canal to provide a navigable link to the sea, running through the west of Galway city and west of the Corrib river. The canal, which was completed in 1852, was named after the Earl of Eglinton. It remained navigable until 1954, when its connection to the sea was compromised by the replacement of two swing bridges with low-level road bridges. The nineteenth century was the heyday for traffic on the lake, and a good many of the quays still in existence were built during that century. Cargoes of stone, turf, crops and livestock were carried by water, rather than on the poorly developed road network. As part of a drainage scheme carried out between 1846 and 1850, a navigable channel suitable for paddle steamers was created on Lough Corrib, and regular passenger services from Cong, at the northern end of the lake, to Galway city began. The completion of the Galway–Clifden railway line in 1895 sounded the death knell for this service, and by 1918 passenger traffic had all but ceased.

Comparisons across the years

At first sight, Lough Corrib may appear to be a daunting challenge as a cycling circuit. With over 200 km of shoreline and a vast array of interesting sites to explore, a single-day circuit could not be contemplated, even if one were to resort to motorised transport. Indeed, a week spent exploring the shores of the three connected lakes of Corrib, Mask and Carra would be a worthwhile and rewarding expedition. One of the more interesting aspects of cycling around Lough Corrib is the opportunity it affords to compare the area with how it presented itself to Sir William Wilde more than a century and a half ago. Wilde's ancestral home was at Moytura, near Cong in County Mayo, and he wrote what is regarded as the definitive work on Lough Corrib, entitled Lough Corrib: Its Shores and Islands. Since its publication in 1867, the book has delighted many readers, although available copies have become rare in recent years. Fortunately, since it was reprinted by Kevin Duffy in 2002, a new generation of readers can be entertained by this detailed study of the lake and its surrounding area.

I have chosen the historic village of Cong as the starting point for the circuit and have opted for an anticlockwise circuit, so that the more onerous and exposed north-western element of the circuit is undertaken with fresh limbs and eyes. My choice of Cong is partly in deference to Wilde but also because it has much of interest in the locality. Indeed, an entire day spent exploring Cong and its hinterland would be rewarding in itself. The Lough Corrib region is one of the most cycle-friendly areas that I have visited on my travels: it has a labyrinthine maze of quiet country roads and lanes supplemented by rough forested trails that enables passage from one site of interest to another without having recourse to the main roads.

Cong

Cong's fame is disproportionate to its size. Nestled amidst thick woodland and surrounded by streams and rivers, this small village boasts a heritage that has an abundance of historical content, and presents it well. The Cong Canal has already been covered in the circuit around Lough Mask, but there is plenty more to see in the village and surrounding area. The village name is derived from the Irish word 'Cunga', which translates as 'a narrow strip of land' and describes its location on the isthmus between Lough Corrib and Lough Mask. In ancient times, it was called 'Cunga Feichin', in recognition of its patron saint, Feichin, who is better known for his association with Fore, County Westmeath (see page 305).

Cong Abbey

The village is dominated by the ruins of Cong Abbey, which lie to the south-west on the fringes of the grounds of Ashford Castle. The ruins are much changed from when Sir William Wilde visited them in the 1860s, and their present state is considerably different from the illustration by Samuel Lover RHA (1797–1868) included in Wilde's book, which depicts the abbey some forty years prior to the book's publication in 1867. It is generally accepted that the abbey dates back to the early twelfth century and was built on a site previously occupied by a monastery founded in the seventh century, presumably by St Feichin. It is believed that it was built by Turlough O'Conor, king of Connacht and high king of Ireland, and became the home of the Canons Regular of St Augustine, the same order that later occupied Ballintubber Abbey, on the shores of Lough Carra to the north. Much of the original structure has succumbed to the cannibalistic tendencies of earlier times, when local builders were no respecters of history and heritage, but what is left is now well presented. Some fine examples of stone carvings have survived on the site and may be seen around the preserved doorways and windows. In the nineteenth century, due to the intercession of Sir Benjamin Guinness, the owner of Ashford Castle, in whose grounds the abbey is sited, the first efforts at restoration were made. According to Wilde, a local craftsman called Peter Foy was employed to restore some of the carvings, so some of the present display is the work of this man. The main access point is now through the chancel leading into the sacristy, from which you can pass through an elaborate doorway to the reconstructed cloister, one of the abbey's more impressive features. Remnants of the Chapter House, including some fine windows and the refectory, can also be seen.

The abbey was suppressed during the reign of Henry VIII and became the property of the King family, whom we have already encountered at Boyle, near the shores of Lough Key. It subsequently passed through the hands of various families, finally coming into possession of the Brownes, who built Ashford Castle.

A pathway flanked by massive yew trees leads from the abbey to the river, where you will find an interesting small stone building at the end of a short jetty built out into the river. This is the monks' Fishing House, which dates back to the fifteenth or sixteenth centuries. The building was used to provide a little comfort to those monks who were given the fishing detail, and perhaps to make their task a little easier. The house is sited on a small platform of stones with an arch, which allows the water of the river to flow underneath the floor of the house. There is a hole in the floor through which the monks could drop lines or a net. The existence of a fireplace suggests that this was a year-round activity for the monks. Wilde relates that when fish were caught in the monks' net, the net was connected to a wire that ran all the way to the monastery kitchen, where it was attached to a bell that would sound at the arrival of a fresh catch.

A bridge across the river leads to Cong Wood, a dense forest of oak, horse chestnut, alder and cedar trees, with delightful trails suitable for off-road cycling. On the path to the wood, you will come across two arched doorways that have interesting sculptured heads etched on to their keystones. The one nearest the abbey is of an ecclesiastic whose nose is unfortunately missing. The other is closer to the wood and is of a finely featured royal person complete with his crown.

Back at the abbey, there is a small modern rectangular church building squeezed into the fringes of the abbey grounds. This building was erected in the early 1970s and was recently refurbished. An attempt has been made to marry this structure with its historical surroundings, via the use of grey stone. However, its ugly grey raised roof over the altar area lets it down and is not in keeping with the surrounds. This leads me to wonder why it was felt necessary to squeeze the church into these grounds. Surely it would have been better to leave the abbey ruins unadulterated by this modern interloper.

The Cross of Cong

One of Cong's most celebrated treasures is the Cross of Cong, a fine ornate processional cross that is now housed in the National Museum in Dublin. The cross was made for Turlough O'Connor as a shrine for a relic of the true cross of Christ that had been sent to him from Rome in 1123. It was made in County Roscommon by Maelsiu Mac Bradan O'Echan, who etched his name on one of its sides and was probably brought to Cong by one of the early abbots of Cong Abbey.

The basic fabric of the cross is oak, which was plated with bronze and silver and decorated with gold and jewels. The relic was housed under a large polished crystal positioned in the centre. Wilde reported that after a number of centuries, its relic had disappeared, and the cross was purchased in 1839 by a Professor McCullagh, who presented it to the Royal Irish Academy. More light on its history is shed by Peter Murray in his book George Petrie (1790–1866): The Rediscovery of Ireland's Past. He describes a visit made by Petrie in 1822 to a Reverend Prendergast PP, who advised Petrie that he had found the cross in an oaken chest that had been kept in a cottage in the village some years earlier. He believed that it had remained hidden there since the Reformation, or at least since the Irish Rebellion of 1641. According to Murray's account, the cross was procured for the Royal Irish Academy by George Smith, at the behest of Professor McCullagh, for the sum of £100. The whereabouts of the relic for which it was made are unknown.

The Quiet Man

No visit to Cong would be complete without reference to a famous Hollywood film that was made in the area in 1951. The Quiet Man, which was directed by John Ford and starred John Wayne, Maureen O'Hara and Barry Fitzgerald, painted a stylised version of life in rural Ireland. The film tells the story of retired prizefighter Seán Thornton, who returns to his Irish roots and falls in love with a fiery local beauty called Kate Danaher. It was the first Technicolor film to be shown in Ireland and acted as a showcase for rural Ireland, subsequently attracting many visitors to the country. The Quiet Man Heritage Cottage replicates some of the sets used in the filming. The Church of Ireland church that lies in the grounds of Ashford Castle was also used in the film.

Ashford Castle

The road opposite Cong's tourist office, which runs by the side of the abbey, leads to a side entrance to Ashford Castle. The main entrance lies off the R346, about 1 km east of the village. The castle is now a thriving upmarket hotel that benefits from a stunning location on the shores of Lough Corrib, amidst extensive mature woodlands. I was somewhat surprised to learn that cyclists are welcome to cycle through the magnificent grounds of the demesne. Indeed, the hotel accommodates cycling tourists and encourages them to avail of the many quiet wooded trails, leading away from the hotel, which will enable them to tour the district around Cong and Clonbur with only rare interruptions from motorised transport.

The castle was originally built in the early eighteenth century by the Browne family but is now a dramatically different structure from its original French-chateau form. It became the centrepiece of a fine shooting estate and in the 1850s was purchased from Lord Oranmore by Sir Benjamin Guinness (1798–1868), a grandson of the founder of the brewing dynasty. To Sir Benjamin, it was initially perhaps a public demonstration of his wealth and status, but it quickly became a favourite family retreat. Sir Benjamin's son Arthur Edward Guinness (1840–1915) inherited the estate on his father's death, and he and his wife, Olivia, were to leave the greatest mark on the property. Having travelled extensively, they set about modifying and extending the castle using a mélange of styles and incorporating French-renaissance roofs, Gothic crenellated turrets and a fanciful Tuscan campanile. Arthur was a great lover of trees and carried out extensive afforestation throughout the estate. The oak tree was his favourite, and two mature oaks grow close to the castle.

Arthur was hailed as a great philanthropist. In 1874, he became a Member of Parliament for the Conservative Party and two years later decided to cut all ties with the family brewing business, receiving the massive sum of £680,000 to relinquish his interests. He devoted a considerable portion of his fortune to the benefit of the Dublin public. A fine example was the rebuilding of the Coombe Hospital in 1877. He was a pioneer of slum clearance and the building of new homes for Dublin's labouring classes. He remained a Member of Parliament until he was ennobled in 1880, becoming Baron Ardilaun – using the name of a large island close to his Lough Corrib retreat.

Lord Ardilaun and the Irish Land League

Lord and Lady Ardilaun's Ashford estate lay in the midst of an area that was central to the Land Wars of 1879 to 1882. Ever since the Great Famine of the 1840s, Ireland's fragile rural economy was subjected to bouts of poor harvests that once again threatened the lives of its poor tenant farmers. An economic crisis caused by poor yields between 1877 and 1879 gave birth to a new organisation called the Irish Land League, whose aim was to improve the lot of the Irish peasantry through fighting for fair rents and fixity of tenure. Michael Davitt (1846–1906) was the founder of the organisation, and Charles Stewart Parnell (1846–91) agreed to be its first president. While the League espoused non-violent methods, such as ostracism and boycotting, assaults on landlords and their agents became widespread. The level of serious and violent disruption was higher in County Galway than in any other county except for County Kerry. The League eventually called for a suspension in the payment of rents in an effort to force the government to pass reforming legislation. The Land Act of 1881 went some way towards meeting the demands of the protestors. It provided for dual ownership between landlord and tenant and gave tenants the three F's: fair rent, freedom of sale and fixity of tenure.

Lord Ardilaun was regarded by his tenants as a benign landlord, but this did not stop him from siding with his fellow landlords. Despite the Guinness name, which had come to be regarded as truly Irish rather than Anglo-Irish, and the respect in which he was held by his tenants, his estate was not immune from attack, as we have seen with the murder of his agent (see page 223). Furthermore, he did little to help his cause by being one of the few to come to the aid of Captain Charles Boycott when, at the behest of Parnell, the latter had been ostracised for refusing to reduce rents for the tenants of Lord Erne, who owned land on the shores of Lough Mask. Lord Ardilaun used his own steamer to transport provisions for delivery to Boycott and even mustered a force from among his own estate employees to escort the goods to Lough Mask House, where Boycott lived.

Lord and Lady Ardilaun were childless, and when Lord Ardilaun died in 1915 the barony came to an end. He left the Ashford demesne to his brother Edward Cecil Guinness (1847–1927), the First Baron Iveagh. The castle was eventually sold off and the estate was bequeathed to the nation.

Before leaving Cong and Ashford Castle, it is useful to note that boat tours to Inchagoill Island can be taken either from the quay beside the castle or from Lisloughrey quay, 2 km east of Cong off the R346. These services are normally only available from April to October.

To return to the lake circuit you will need to track back to the pillars that you passed as you made your way from Cong to the castle. The road to the left after the pillars leads to the gates of the estate's old school, and this is followed by a delightful tree-lined avenue. The first turn left along this avenue leads towards the lakeshore once again. It doesn't matter if you miss that turn, as the avenue eventually emerges at the entrance to Ardnageeha Wood, where a turn left allows you to rejoin the road you missed. A turn off the road leads to a shore-access facility at Cappacorcoge that you may wish to visit. The tight roads along this route allow you to track the shoreline on an undulating course that yields intermittent views of the lake. A remarkable feature of the landscape is the artistry of the walls separating the small fields on the hilly terrain rising from the lakeshore. As you approach the R345, the views of the lake improve and the scattering of islands wedged between the shore and Doorus peninsula become visible.

With the slopes of Benlevy rising to your right, you descend on a winding road towards Cornamona. You may notice nets over the road between trees on opposite sides. When I enquired from locals as to the purpose of these nets, I was told that they were to allow safe passage for the squirrels.

Doorus peninsula

Cornamona, which translates as 'the Rough Field of the Bog', stands at the base of the Doorus peninsula and is dissected by the Dooghta river (also known as the Cornamona river), which drains the mountains of 'Joyce's Country' into Lough Corrib. At 6.5 km in length, Doorus peninsula is a worthwhile extension to the circuit, if only for the close view of Inchagoill Island and the dramatic views of the mountains it offers as you make the return journey. There are various shore-access points along the peninsula, and where the metalled surface ends there is a right of way through the gate to the shoreline at the tip of the peninsula, where there is a fine view across the broad expanse of the upper lough and the dominant Inchagoill Island.

Inchagoill Island

The modern Irish version of this island's name is Inis an Ghaill, which means 'Island of the Foreigner'. Wilde provides an extended version – 'Inis an ghoill Craibhtheach' – which translates as 'the Island of the

Devout Foreigner' and refers to an unknown holy man who lived on the island in ancient times. The island was bought over by Sir Benjamin Guinness after he had purchased Ashford Castle. He arranged for the enclosure of the island's graveyard and the partial restoration of the larger of the two churches, of different dates, to be found there. The more northerly and plainer of the two churches is known as 'Teampall Phadraig', or 'St Patrick's Church', while the larger and more modern church is called 'Teampall na Naoimh', or 'the Saint's Church'. The two churches are connected by a flagstone path.

Lying near the older church is an unusual monumental stone known as Lughaedon's Pillar. The stone is just over half a metre tall and has two crosses cut into the face of three of its sides, while the remaining side contains a single cross. There is an inscription on its east face which reads 'Lie Lughaedon MaccI Menueh' and identifies the stone as belonging to Lughaedon, the son of Meneuh (or perhaps Imenueh, depending on where the 'I' is located). According to Wilde, Lughaedon was St Patrick's nephew.

As mentioned earlier, the island can be reached using the seasonal boating services available from Ashford Castle or Cong, and also from the western shore at Oughterard.

Maam

Returning to the mainland, there is a fairly steep climb away from Cornamona, followed by a prolonged descent towards the Connemara portion of the lake. As you descend, there is a picturesque view of the isolated ruins of Hen's Castle, perched on a rock close to the lakeshore. According to Wilde, this castle is one of oldest mortared castles in Ireland and is attributed to both the O'Conors and the O'Flahertys. Wilde provides a number of fanciful theories as to the origins of the name, the most sensible of which states that it was named after the widow of a man known as 'Cock O'Flaherty', who became known as 'the Hen' after she had successfully defended the castle from raiders. It is reputed to have been the last Irish castle to have remained in Irish hands: it fell to Cromwellian forces in 1653. The Guinness family took control of the castle in the 1800s, but not before a good deal of its stone had been cannibalised for use in the construction of local cottages.

An undulating course flanked by the lake on one side and the imposing Lugacurry and Knocknagussy peaks on the other carries you towards the village of Maam. The lake has reached its most northwesterly point and terminates south of the village at the estuaries of the Bealanabrack and Failmore rivers. Wilde describes the Bealanabrack river as 'the largest stream in Ireland for its length'. It is an attractive river, amply filled from the slopes of the Connemara and 'Joyce's Country' mountains. The name is translated from Irish as 'the River of the Trout's Mouth'.

The Connemara landscape

Just before you reach the village, it is necessary to turn left over Maam Bridge, which is adjacent to Keane's bar, and set off in the direction of Maam Cross for the start of the journey south along the western shore of the lake. This will take you away from the lakeshore for an extended period – in fact all the way to Oughterard, a distance of over 20 km. I do not regard the journey inland as a betrayal of the principles of lakeshore exploration, as the landscape you will be travelling through is dotted with some quite substantial lakes along the route. In addition, you will have the opportunity to sample some of the spectacular Connemara scenery at first hand. The more adventurous, or 'lakeshore purists', might like to pursue the trail of the Western Way, which in fact keeps to the lakeshore and can be accessed by taking the first turn left after crossing the bridge over the Failmore river. This small road is metalled for the first kilometre and then deteriorates, becoming a rough path populated by grazing sheep. It is not recommended during the winter months but presents a passable challenge for a well-equipped off-road bicycle in drier conditions.

The road to Maam Cross is sandwiched between two peaks: Lackavrea to your left and Shannaunnafeola to your right. According to Wilde, the Irish name for Lackavrea, 'Leic Aimhréidh', translates as 'the Tortuous Slate' or 'Flag' and describes the slate-like appearance of the quartzite rock on the surface of this mountain. He is less definite about the origins of the name of the opposite peak. A loose translation is 'the Old Flesh of Deer or Cows', and it may be derived from the animals that used to dwell on its slopes. Sheep are its only current inhabitants. Under the lee of Lackavrea lies Maumwee Lough, which appears to have a crannóg site, although none is indicated on the map. The road past the more substantial Loughanillaun is fairly exposed and in windy conditions presents quite a challenge. As the landscape changes, so does the light, and it is no wonder that this area has attracted so many artists over the years. Even on the dullest of days, the variations in colour are inspiring. There is no single colour attaching to any of the landsape features. Bogs are a blend of brown, green, yellow and purple. Quartzite mountain summits shimmer in silver-grey, which contrasts with the dark black and coppery shadows created by the varying contours of their slopes. Meanwhile, the succession of small lakes reflects and amplifies the wild, multi-hued environment. This is truly an artist's paradise.

Maam Cross

Maam Cross is known as 'the Piccadilly of Connemara' and plays host to a variety of annual events that commemorate the location as a meeting place for isolated communities down through the years. Each October, it is the venue for a pony fair. One of the animals traded is the distinctive Connemara pony, a hardy animal that is thought to be a product of local and Spanish stock. The fair, which now attracts upwards of 15,000 people each year, developed from the tradition of locals selling their surplus produce, including potatoes, butter, mutton, beef and wool, to people visiting the area. There is also a Bogman's Ball held every February to commemorate the customs of the bogmen and their rapidly disappearing way of life. The festival began in 1959 to celebrate the end of the annual campaign to save turf for fuel. Festivities centre around demonstrations of bogmen's customs, including cooking steaks on shovels on open fires in the traditional bogman manner.

The road to Oughterard skirts past several lakes that are distinctive because of ther long, narrow profile. Both Ardderry Lough, which is sited beside Maam Cross, and Lough Bofin, further along the route, stretch out for over 2.5 km but are mostly only a couple of hundred metres wide. The lakes are nearly all connected by streams and rivers, which ultimately contribute to the Owenriff river, a significant tributary of Lough Corrib on its western shore. Accordingly, the connection with the bigger lake is, in a way, being maintained.

Glengowla mines

The road starts to turn again towards Lough Corrib just after passing Lough Agraffard. A short ride further on is the entrance to Glengowla silver and lead mine. This mine was abandoned in 1865 but has been reopened as a show mine, attracting visitors who wish to witness at first hand nineteenth-century methods of ore extraction and processing. When the mine was operational, the ore was shipped to Galway on the waters of Lough Corrib. Tours bring visitors underground through large marble chambers and caverns, where rare and beautiful rock formations can be inspected, including specimens of Connemara's famous green marble. A heritage and visitor centre provide the background information you will need to appreciate this unusual attraction. The mines are open daily from March to November.

Oughterard

Oughterard is the first large town encountered on the western shore. For those who chose the inland route instead of the Western Way course along the lakeshore, there is the opportunity to backtrack northwards to Doon Rock, so that you miss as little of the Lough Corrib shoreline as possible. The 14 km route from Oughterard to Doon is one of the highlights of the circuit in terms of scenery and cycling pleasure. The road is narrow but has little traffic, particularly as you approach Doon Rock. The route skirts around the base of Knockbrack Mountain and Derroura Wood and has a captivating final 3 km, where the road runs right by the lakeshore, with views across to Doorus peninsula. At the parking areas opposite Doon Rock, you will encounter a peaceful scene where the only sounds are the water lapping the shore and the odd snatch of birdsong. It is a beautiful, serene and enchanting place. An indistinct path leads down from the parking area to the lakeshore, where a crumbling, grass-covered jetty is an indication of how few people visit this jewel of Lough Corrib, particularly during the winter months. Doon Rock is densely wooded, lying at the tip of a ridge of land called Drumsnauv ,and is strategically placed at the channel that leads into the Connemara portion of the lake.

Oughterard is described in the tourist literature as 'the gateway to Connemara'. In our case, it is where we bid farewell to that rugged countryside and head south through an area described by Wilde as 'desolate' and a 'low and sterile district'. Notwithstanding Wilde's reservations, it is an area of considerable appeal to cyclists, as it is largely flat, with quiet country roads that run fairly close to the lakeshore. It also hosts a number of sites of historical importance, including the hugely impressive Aughnanure Castle, which was at one time the principal residence of the once-powerful O'Flaherty clan.

Aughnanure Castle

To find the castle, you should travel 2 km south of Oughterard on the N59 and turn left by the golf club. The castle is located about 1 km to the east, on the south bank of the Drimneen river. The castle's name is derived from the Irish 'Achadh na n-Iubhar', which translates as 'the Field of the Yews' and relates to a forest of yew trees which grew in the area in ancient times. The imposing six-storey tower house was built in the early sixteenth century. There is some disagreement as to whether it was the De Burgos or the O'Flahertys who actually started the building project, but it is accepted that it was the O'Flahertys who made the greatest mark on the property.

By the sixteenth century, the O'Flahertys were in fact a divided clan. The territory known as Iar-Connacht (west Connacht) was under the control of Murrough na Doe O'Flaherty, who had declared his allegiance to the English Crown and was rewarded with the possession of Aughnanure Castle – or what was left of it, as it had been partially destroyed in 1569 by Edward Fitton, Lord President of Connacht. Murrough restored the castle and used it as his main residence. The rightful chief of the O'Flahertys was Donal an Chogaidh, the husband of Grace O'Malley, who later became known as the celebrated pirate queen Gráinne Uaile. Donal ruled Connacht but was killed in 1570 while raiding Galway city. His wife and son refused to accept allegiance to the English Crown or to Murrough na Doe O'Flaherty as the clan chieftain.

In 1618, James I granted the castle to Hugh O'Flaherty. His son Roderic was born there in 1629 and went on to find fame as a scholar and writer. In 1684, he became the first Irishman to write a chorographical account (a detailed account of the topographical and geographical features) of a part of Ireland, describing the natural features of his homeland of west Connacht. (Wilde makes frequent references to this seminal work.) In the middle of the seventeenth century, Cromwellian forces dispossessed the O'Flahertys and the castle was used as an outer defence for Galway city, to prevent attack from the west via Lough Corrib. After the Restoration, Roderic O'Flaherty petitioned for the return of the castle, but even though he was successful in his pleas, as a Catholic, he and his family suffered badly after the enactment of the Penal Laws of 1695 to 1709, and he died in poverty in 1717. After his death, the castle fell into ruin.

It is not often that I have found a ruin in much better condition than Wilde found it over 150 years ago, but I am pleased that the efforts of State agencies in recent years have restored the tower house to its former glory. It is a pleasure to visit a piece of our heritage that has been rescued from the mundane fate of being used as a dairy farm – as was the case in Wilde's time. Apart from the repairs to the tower house, the remnants of the walls of the inner and outer bawns (fortified enclosures) have been carefully preserved. Located along the inner wall, close to the tower, is a fine small round house with a corbelled roof. The dark waters of the Drimneen river form a natural defence on one side of the castle, and a part of the castle's banqueting hall collapsed into one of the underground channels of this river even before Wilde's time.

The western shoreline of the Middle Lough

Resuming a southerly course, you will need to consult the map closely to avoid ending up back on the N59, as you can quickly lose your bearings on the tight, meandering country lanes in the area near Aughnanure Castle. You should eventually emerge on to a relatively straight road that will take you fairly close to the lakeshore for an extended period, all the way to the small village of Carrowmoreknock and then Knockferry. All along this stretch the lake is narrowing, until it reaches its narrowest width, of only several hundred metres, between Knockferry on the western shore and Kilbeg to the east. In ancient times, this was the chief point of passage between eastern and western Galway. Wilde reported on proposals in his day for a causeway or bridge to be built across the lake at this point. The proposals never bore fruit. It remains a logical point of connection and would cut out 50 km-long road routes between Galway and Connemara.

In recent years, there have been proposals by Shannon Ferries

Limited to use a purpose-built cable-operated car ferry with a capacity of twenty-four cars to link the two shores. The ferry was to have been similar to one currently operating on Lake Windermere in England and would have been the first of its type to operate in Ireland. The proposed journey time was seven minutes. Planning permission for the ferry proposal was denied in April 2006, on the grounds that insufficient information had been submitted on the impact the ferry would have on the lake's freshwater environment, but scope was left for a revised application at a later date.

Pushing ahead, you can perhaps appreciate Wilde's rather depressing sentiments about this area as you travel through a flat and rocky limestone region. To lighten the mood, you can focus on some of the unusual place names of townlands you pass through on your journey. Just beside Knockferry is Burnthouse – not to be confused with Ballydotia, which you will encounter later. The translation of 'Ballydotia' from its Irish form is 'Burnt Village'. Further along, you have Wormhole, located by some small lakes, which may well have given rise to the name. A small loop leads past Gortmore, where you can still spot evidence of the lead-mine operations that used to exist there. This is one of about six small lead mines that were operated in the area in the middle of the nineteenth century by the local landlords, the O'Flahertys. As with Glengowla, the ore from Gortmore was shipped to Galway city via Lough Corrib.

As you progress back towards the N59, there are several left turns that lead off to the lakeshore that you may wish to explore. One small diversion worth undertaking brings you to the ruins of Tullokyne Castle, also known as 'the Hag's Castle', from the Irish 'Caisleán na Cailliaghe'. This was an early O'Conor castle whose history appears to have been lost in the mists of time. Further on, it is possible to bypass Moycullen altogether by using the route that runs past Ballyquirke Lough, south of the village, to link up with the N59 further south. This is the area that we mentioned in the introductory comments about Lough Corrib. Manannán was slain by Uillin, hence the name 'Maigh Cuillinn', 'the Plain or Field of Uillin'. Just after you cross the canal that drains Ballyquirke Lough into Lough Corrib, you will find Patrick's Holy Well to your right, followed by the ruins of a small church called Teampall Beg.

Emerging on to the N59 at Addragool, you have arrived at a point almost parallel to Lough Corrib's southern extremity, where it empties into the River Corrib. It is not possible to get close to the lake's relatively narrow southern shoreline, and you will have to continue on the busy main road as it follows the course of the River Corrib towards Galway city. Follow the directions for the N6, which will carry you across the River Corrib, and as soon as you cross the river, you have the chance to leave the main road by taking the first turn left in the direction of Menlough.

Menlough Castle

In the middle of the nineteenth century, the village of Menlough was a populous place, with a pre-Famine figure of 1,100 inhabitants. Wilde reports that by 1861 this had fallen to 682. He also remarks that Menlough Castle was 'one of the handsomest of the inhabited old castles of Ireland'. Much has changed in the intervening years. The village is a shadow of its former self and the castle is now in ruins and well hidden from the road. There is no signpost or other indication to the casual passerby of its existence, other than a ruined archway and turret that was once the grand entrance to the Menlough demesne. The castle is to be found down the laneway that leads from the ruined arch and is sited by the river bank, at the base of a private field now inhabited by a number of horses. Its setting is spectacular – on a wide, gentle bend in the river – and it was once a majestic structure with its own private harbour. Alas, it is now a shadow of its former self: its present-day function appears to be to provide shelter to livestock. The castle was owned by the Blakes, one of the former merchant families of Galway and one of its famous 'tribes'. The Blakes were in residence right up to 1910, and it is remarkable how much the structure has deteriorated in the intervening period. The ivy coat that existed in Wilde's time remains and is probably the reason why the ruins are still standing. To the rear of the castle is a small wood planted on a hill that helps to obscure a land-side view of the ruin.

A good view of the Friar's Cut, mentioned earlier, can be obtained from the road that leads down to Menlough graveyard, which can be accessed to the left just after the arched entrance to Menlough Castle.

A very narrow winding road takes you away from Menlough, passing by a number of small but delightful thatched cottages – of which there would have been many more when the village was a thriving satellite of Galway. About 2 km from the village, the wide expanse of the lower lough, fringed with dense reed beds, comes into view. The views improve after you pass an extensive quarry and the road starts to rise. The other striking impression as you make your way towards the N84 is the broad expanse of Galway's eastern plains.

The N84 is the main road between Galway and Castlebar and is extremely busy. There is no option but to use it until you reach the small village of Clonboo. It is, however, straight and flat, and swift progress can be made amidst the poor-quality marshland that is a feature of the immediate eastern shores of Lough Corrib. As you track northwards to Clonboo, there are good views across the lake to the ridge of hills that stretch from Galway to Oughterard. Two rivers are crossed in quick succession. The first is the Clare river, which is a fairly wide watercourse that has been partly canalised. It is one of the more important sources of water for Lough Corrib and also hosts a safe spawning ground for the lake's salmon stocks. The second is the Cregg river.

Annaghdown Castle

Almost imperceptibly, the N84 has been taking you away from the shores of Lough Corrib towards better-quality land. This is recognised at Clonboo, whose name means 'Fertile Land of the Cow'. At Regan's pub at Clonboo, you can at last turn left off the busy main road and once again make for the lakeshore along a quieter course that leads to Annaghdown. The first sight you see as you approach this area is the superbly restored Annaghdown Castle, whose condition is considerably improved from that illustrated by Wilde in the middle of the nineteenth century. The castle is believed to have been a de Burgo tower house originally. According to an article by Davin O'Dwyer in the Weekend supplement of the Irish Times of 29 July 2006, the restoration was undertaken by Jessica Cooke and her husband, movie producer Seán Faughnan, from 1997 to 2006. The building is now in use as a family residence. The article reported that Ms Cooke inherited the castle from her father, who owned it for many years. Restoration is obviously in the bloodline: Mr Cooke was a structural engineer who had been involved in the restoration of Dublin Castle. From the evidence unearthed during the restoration, the present occupiers commented to Mr O'Dwyer that they are of the opinion that the castle had been uninhabited since the late seventeenth century. There is no public access to the castle or its grounds.

Annaghdown was the first stop for steamer traffic from Galway and at one time had a vibrant village. When Wilde passed through the village in the mid-nineteenth century, he reported that there was only one inhabited house remaining – occupied by the man who attended the steamer. Two factors contributed to the decline of the village. The first was the tragic drowning of nineteen of its inhabitants, eleven men and eight women, on 4 September 1828 as they were making their way with their animals to the Galway Fair aboard an old boat that sank near Menlough. The names of the victims are listed on a memorial erected in modern times by the Annaghdown Angling Club at the pier from which they set off on their fateful journey. The tragedy is also commemorated in an Irish poem that is familiar to most who have passed through the Irish education system: 'Eanach Chuain'. The poem is normally attributed to the celebrated Irish poet Antoine Ó Rafteirí, but it may have been composed by another poet called O'Ceallaigh. The second tragedy to strike the village was of course the Great Famine, which resulted in the village becoming deserted. Fortunately, it has since recovered, and its modern school is evidence of a now-thriving community.

Annaghdown Priory and churches

The pier is an excellent vantage point from which to admire the middle portion of the lake and the nearby castle. On the access road to the pier, there is an interesting ecclesiastical ruin, one of a number in the immediate vicinity – none of which are signposted or otherwise advertised to the passerby. It is said that St Brendan was the first to found a monastery in this area, in the sixth century, leaving it under the control of his sister, an Augustinian nun. The ruin nearest the pier is of a priory that dates back to the late twelfth century and is believed to have been occupied by nuns belonging to the Order of Aroacea. The ruin is somewhat diminished since Wilde's time but fortunately any further decline has been prevented by recent preservation works. Sited to the north of the priory, located in an extensive graveyard and in a more elevated position, are more ruins. These ruins are better accessed via the main entrance to the graveyard, which is on the main village road. Here you will find the remnants of two churches, one of which dates back to the fifteenth century. This is significantly larger than a normal parish church and is referred to as a cathedral church. Within the walls of this church, the restorers have gathered all the stones that at one stage must have formed the structural walls of this building and laid them out in a tomblike edifice.

Before leaving Annaghdown, I think it would be wise to clear up some of the confusion regarding the title of the poem composed in memory of those who lost their lives in the boating tragedy referred to earlier. What is the link between the village's name and the title of the poem? Annaghdown is an anglicised version of the Irish 'Eanach Dhúin', which translates as 'the Fortress of the Bog'. The village is within an area that is close to a parish associated with St Coona or St Cuanna. It is possible that the poet was referring to the larger parish area, as some of the victims came not from the village itself but from the neighbouring townlands.

Ballinduff bay and castle

The road away from Annaghdown leads back to the N84 along the southern shore of Ballinduff bay, the largest inlet on the eastern shore. This is a busy and sometimes quite narrow main road that in places leaves little room for cyclists, so care is necessary. Along the way, you will see the ruins of another two castles. Firstly, Ballinduff Castle appears to the west. This belonged to the Skerritts, another of the merchant families of Galway who vacated that city for a lakeshore residence. The ruins are in a particularly poor state of repair and have deteriorated considerably more than most of the structures of similar vintage along the lakeshore. The next set of ruins is to the right of the road at Cloughanover, where you have the chance to leave the N84, turning left. The Irish name for the townland is 'Cloch an Uabhair', which means 'Stone of Pride' and, according to Wilde, relates to a stone that lies near the castle. Once you have turned left off the main road, it is an almost direct route back to the lakeshore. However, you should look out for the signpost pointing to Kilbeg pier so that you will have the opportunity to inspect the narrowest point of Lough Corrib from its eastern shore and look across to Knockferry, which you visited earlier on the circuit.

Cargin Castle

Moving along the lakeshore, you will come across Cargin Castle, which is heavily concealed from the road. Originally, this was a hall house with a small rectangular tower at its east angle. It was bought in the 1960s by Christopher Murphy, who lives in London. As the castle is in densely wooded private grounds, there is no opportunity to inspect it at close quarters, but from the little you can see of it, it appears to have been considerably restored and modified in recent years. Cydagh House is similarly well concealed: all you get to see is the stone wall of the wooded demesne.

Annaghkeen Castle

Annaghkeen Castle, located 2 km to the north, is more rewarding. As you descend to the lakeshore, the ruins of two fine buildings come into view. They rest close together in a generously proportioned meadow with stone-walled boundaries. It is quite obvious that the two ruins are of a different age. The older building is almost square, and Wilde believed that it may have been the oldest castle on the shores of Lough Corrib. It is finely built and, unlike other castles in the region, its stonework does not appear to have been cannibalised for alternative uses, even for the building of the later structure on its southern flank. The earlier structure, which probably dates back to at least the fourteenth century, passed through various hands, including those of the Lynches and the Burkes. Wilde reported the more modern structure to the south as being unfinished and belonging to a Mr O'Flaherty – which makes me wonder if it was ever occupied.

Inchiquin Island

The trail along the lakeshore north of Annaghkeen is a labyrinth of tight winding roads bounded by low stone walls. It is an area of few sites of historical significance but is pretty nonetheless in a rugged sort of way. The route follows the jagged contours of the lakeshore and is peppered with many lanes, leading right down to the shore, that you may wish to explore at your leisure. One diversion worth taking off the circuit is to Inchiquin Island, which can be reached via a man-made causeway. The inhabitants of the island persuaded the authorities to build the causeway by withholding taxes until it was complete. The causeway is very narrow, with just enough room to accommodate a single car. The road on the island is confined to its eastern shore, but because the island is low-lying there are fine views of the lake from this side. The island is the largest in Lough Corrib, covering 93 hectares. Its name means 'the Island of the Sons of Con', who was king of Ireland in the second century. Wilde reported that there was scarcely a vestige remaining of the ruins of ecclesiastical structures built on the island by St Brendan and St Meldan in the sixth century. Whatever existed is now no longer visible.

Returning to the circuit, the tight roads allow you to negotiate a passage through drumlin hills, which at times inhibit the view of the lake. The very narrow Gortbrack Bridge over the Black river greets your return to County Mayo, although there is no change in either the landscape or the road surface. Wilde described the area immediately north of the Black river as 'dreary', as there was scarcely a house or tree to break the monotony of the scenery. Not too much has changed in the intervening period, and it remains a sparsely populated area. Nonetheless, I would disagree with the great man in his description: from the saddle of a bicycle, it is an agreeable and calming area, with little traffic and frequently pleasing views of the lake.

Inismactreer Island

Another worthwhile extension to the circuit in this area is a visit to Inismactreer Island, which, like Inchiquin Island, is joined to the mainland by a man-made causeway. Once again, the causeway is very narrow, and to add to the dangers there are no walls along a good part of it. As you make your way across the causeway, there is a sensation that you are in fact travelling on the water, and you can appreciate the attractions of exploring the lake by boat. You can also appreciate how dangerous the lake is for boats, as there are lots of jagged and pock-marked limestone rocks peeping out of the water. The views from the causeway and from the island itself are enthralling, particularly after you have crossed the little bridge at the island end. To the north are the wooded shores around Ashford Castle and Cong. Along the roadway that goes to the southern tip of the island, you will come across the island's Millennium Stone, on which is inscribed the island's original Irish name, 'Inis Mhic an Trír'. Wilde infers a connection between this name and the causeway, which in Wilde's time was only a summer feature. He suggested that the last word may have been a derivative of the Irish word 'tír', meaning 'land', and translates the full name as 'the Island of the Land'. This appears fanciful, and it is more likely that it follows the same line as Inchiquin Island and that the more appropriate translation would be 'the Island of the Sons of Trír', although I can shed no light on who Trír may have been. The extension to Inismactreer Island adds 14 km to your journey.

Cross

There is a back road that leads almost all the way to Cross and avoids the need to use the busier R334 until you are almost at the village. This road can be accessed by taking the second turn left after the signposted turn for Inismactreer Island. This quiet, narrow road will surprise you with its repeated twists and turns.

The village of Cross lies close to the plain of Moytura, which, as mentioned earlier, is, according to Wilde, the location for the First Battle of Magh Tuireadh. Legend has it that the first ever hurling match played in Ireland took place on that plain. The match was between teams from the Firbolgs and the Tuatha Dé Danann, with twenty-seven young men on each team, and was used to settle the war between the two tribes. The rules were obviously quite different from present-day hurling, as all members of the Tuatha Dé Danann team perished during the match, giving victory to the Firbolgs!

Ballymacgibbon cairn

The final leg of the journey to Cong has to be undertaken on the R346. As you cycle west away from Cross, you will see Ballymacgibbon cairn to the left of the road. If you accept Wilde's measurements, this stone cairn has been somewhat diminished by the passage of time. Wilde reports the cairn as being '129 yards [118 m] in circumference, and about 60 feet [18.3 m] high'. Dr Peter Harbison, in his Guide to National and Historic Monuments of Ireland, states that the cairn is about 30 m in diameter and only 7.6 m high. The cairn appears unexcavated and may cover a passage tomb. On the opposite side of the road from the cairn is a ringfort called 'Cathair Phaeter', or 'the Pewter Fort'. Through the heavy growth, you may be able to make out the fort's souterrain.

It is appropriate that as you end the circuit, you pass Sir William Wilde's former residence, Moytura House, which is sited off the R346 about 2 km east of Cong village.

County Westmeath

Lough Ennell
Lough Owel
Lough Derravaragh
Lough Lene

Located in the heartland of Ireland, Westmeath is known as 'the land of lake and legend'. Perhaps the best-known of these legends is that of the Children of Lir, which is associated with Lough Derravaragh. This lake is arguably the finest in the county from a cycling viewpoint as it is surrounded by very quiet country roads and trails. A significant section of the shoreline of Lough Ree runs through Westmeath, but because of its association with the River Shannon it is covered, with its sister lakes along the Shannon system, in a section of their own (see page 106–19). Lough Sheelin might also have been included, as part of its shoreline runs along the Westmeath/Cavan border, but given that it lies mainly in Cavan, I opted to include it with the lakes of that county.

Renowned among the angling community as extremely good fishing centres, the lakes of Westmeath have extensive shallows and high-quality feeder streams and rivers that help to yield good catches of wild brown trout. Lough Ennell is classified as one of a dozen or so wild-brown-trout lakes in Europe. The lake holds the record for the largest trout caught in Ireland, weighing 26 lbs 2ozs. The lucky angler was William Meares, and his catch was recorded on 15 July 1894. The fish is now on display at Belvedere House Visitors Centre, on the eastern shore of the lake. The lakes are also of great significance in terms of wildfowl, with a huge variety of ducks, swans and other nesting

birds such as snipe, woodcock and common terns. The ubiquitous cor-
morant, the bane of many an angler, also graces these lakes.

The cycling circuits outlined below are quite varied. The circuit
around Lough Ennell is relatively flat, with hardly any climbs, while
those around Lough Owel and Lough Derravaragh feature several
descents and ascents, none of which are too arduous. The lengths of
the circuits can also vary depending on whether you choose to divert
off-route to visit the lakeshore by each of the roads and trails depicted on the Ordnance Survey map. If you are using that map, it should
be noted that some of the trails shown in grey merely lead down to
private dwellings and so should be avoided, unless indicated otherwise
here.

River Brosna

Belvedere House

Goose Isd

Blind Isd

Cherry Isd

Jonathan Swift Park

River Brosna

Lough Ennell

(30 kilometres)

LOCATION
Approximately 5 km south of Mullingar off the N52 road

LENGTH
7 km/4.5 miles

WIDTH
4 km/2.5 miles

AREA
1,400 hectares/14.0 km^2

PUBLIC ACCESS
Butler's Bridge, Belvedere, Ladestown, Lilliput and Whitebridge bay

MAP
Ordnance Survey of Ireland Discovery Series Map 41, covering parts of Longford, Meath and Westmeath (ISBN 978-1-901496-30-7), and Map 48, covering parts of Offaly and Westmeath (ISBN 978-1-901496-45-1)

According to a local ballad, Lough Ennell was created by a witch from Connacht with a 'praskeen' of water taken from the River Shannon. She poured the water over a 'hideous bog' and then enhanced the result with islands and fairy gifts. The first verse of the ballad describes the result:

There is a lake in Erin's Isle
And nowhere can one find
In any landscape we have seen
So many charms combined.

These charms have enchanted locals and visitors alike over the centuries and inspired another local ballad, entitled 'The Three Lakes', which includes the following description of the lake:

To the south is the lovely Lough Ennell
Where Malachy's Island does rest.
Rooks and pigeons now call
'Round the old Jealous Wall.
It's a spot that is fifty times blest.

Whatever about the witch's claim to the creation of the lake, there is a link to the River Shannon via the River Brosna, which flows through the lake, entering at Lynn on the northern shore and leaving beside Jonathan Swift Park at Lilliput on the southern shore.

The lake lays claim to royal connections, dating back over a thousand years. At a time when aristocrats lived on lake-island dwellings or crannógs, Maolsechnaill (Malachy) II, high king of Ireland, who ruled in the tenth and eleventh centuries, used the area around Lough Ennell as his centre of operations. He died in 1022 and is buried on the island of Croincha (Cró Inis) on the western shore. With this background, it is not surprising that the area has over the years yielded some interesting treasure. Silver hoards were unearthed, some with silver ingots weighing over half a kilogramme. Other finds include Anglo-Saxon and Viking coins.

A good place to start the Lough Ennell circuit is from the public car-park facility at Whitebridge bay beside Lough Ennell Caravan Park. This is accessed from the N52 by taking the first turn right after the entrance to Belvedere House going in a southerly direction. The small coffee shop at the entrance to the caravan park is open during the summer months.

Returning to the N52, turn right. There is a somewhat inauspicious start to the circuit, as the first part has to be undertaken on this busy road with no view of the lake. This lasts for about 4 km before you take a turn to the right on to a relatively quiet country road that leads towards Kilbeggan. After a further 2.5 km, you come to Newell's Bridge, which carries the road over the weed-clogged River Brosna as it makes its way south from Lough Ennell. Turn right immediately after the bridge in the direction of Castletown Geoghegan. The surrounding area is very open and flat, making it a pleasant cycling route except in high winds. There is still no view of the lake, but following the signpost for the Belvedere Tour and the Jonathan Swift Park will lead you down a quiet lane past a raised bog and a forested area towards the lakeshore at Lilliput, where there is an adventure centre. The lane to the right at the entrance to the adventure centure will bring you to a pleasant picnic area.

Lilliput

Lilliput is reported to have been named after the land that featured in Jonathan Swift's Gulliver's Travels, published in 1726. Swift was a regular visitor to the area, being friendly with George and John Rochfort. He used to stay at Gaulstown House, a Rochfort family residence located off the N6 between Rochfortbridge and Milltownpass, not too far from Belvedere House, which was also owned by the Rochfort family – and which we will come to later. It is said that Swift was inspired to write about the little people of Lilliput after noticing the size of people in the distance as he looked across the broad expanse of Lough Ennell, although this appears to me to be a rather fanciful attribution.

Ballinea and the Royal Canal

Returning to the entrance to Jonathan Swift Park, turn immediately right on to a quiet country road that gradually leads away from the lake through the townland of Dysart. Once again, the terrain is relatively flat, and the lack of traffic makes for a pleasant cycle. After about 4 km, the road meets up with the much busier R391, where you take a turn to the right in the direction of Mullingar. The small village of Ballinea offers the opportunity of moving off-road, taking temporary advantage of the towpath that runs along the Royal Canal. There are two bridges over the canal at Ballinea, as well as a small harbour and boat slipway.

The towpath experience is only a brief one. Once you reach the next bridge, called Bellmount Bridge, cross it to access a delightful tree-shaded lane that brings you to a junction where turning right leads down to the public shore facility at Ladestown, where there are fine views across the lake. There are a number of trails leading away from the car park at Ladestown, either north along the shore or south into a small forest. While the more adventurous off-roaders might be tempted to forge their way along these trails to the next public-access facility at Lynn, which is just over 2 km along the shoreline, this is not to be recommended. The River Brosna enters the lake a short distance west of the Lynn Shore car park and there are no means of crossing the river near the lakeshore. Consequently, there is no alternative but to retrace your steps to the junction with the lane from Bellmount Bridge and continue straight to the next junction, where you bear right. This road carries you over the River Brosna at Butler's Bridge and then on to the busy N52 at Lynn crossroads, where once again you bear right.

Belvedere House

As you make your way back towards your original starting point, you pass the entrance to Belvedere House Gardens and Park, which is well worth a visit. The name 'Belvedere' means 'Beautiful View', and this aptly describes the area around this stately home. Set amidst 160 acres of parkland on the shores of the lake, Belvedere is described as one of the finest of Ireland's historic houses and has been tastefully restored. Opened to the public since April 2000, it was designed in the eighteenth century as a hunting and fishing lodge for Robert Rochfort, Earl of Belvedere (1708–74). The facility is open seven days a week and there is an excellent coffee shop, the Courtyard Café, and a visitors centre on the premises. There is also the opportunity to learn something of the fascinating and often tragic history of the Rochfort family, in particular the extraordinary thirty-one-year imprisonment of Robert Rochfort's second wife, Mary Molesworth. Mary, an actress, was the daughter of Lord Molesworth and married Robert Rochfort in 1736. Her friendliness with Robert's brother Arthur and his wife, who lived at nearby Bellfield House, led to an accusation by Robert in

1743 of an adulterous affair between Arthur and Mary. Robert was awarded £20,000 damages – a huge sum of money in those days – and as a result Arthur spent the rest of his life in a debtor's prison. Mary fared little better: Robert confined her to Gaulstown House and allowed her no contact with the outside world. Her confinement lasted thirty-one years, until the death of her husband.

In his book Irish Ghost Stories, Padraic O'Farrell reported that there have been sightings of a beautiful lady paddling barefooted on the shores of the lake near Belvedere House. The lady is said to be quite animated and gesticulating angrily towards the water. It is said that this lady may be the ghost of Lady Belvedere, the former Mary Molesworth.

The Jealous Wall

As you enter the car park at the end of the long driveway from the entrance to Belvedere, you will immediately be struck by a sizeable ruin known as 'the Jealous Wall'. This is not the remnant of a stately home or other such graceful structure. It was always just a ruin or sham Gothic structure, built by Robert Rochfort to blot out the view of another brother's house, of which he was jealous – hence its name. Sham, or deliberate, ruins were often built in the eighteenth century to decorate the landscape of an estate; they were also known as 'eye-catchers'. The ruin at Belvedere was by far the largest such structure built in Ireland, and it seems an extreme form of profligacy. It is no wonder that Robert Rochfort was deeply in debt when he died.

Another famous past resident of Belvedere was Lieutenant Colonel Charles Howard-Bury (1881–1960), who led the first ever expedition to Mount Everest. He was a keen hunter and angler, and Belvedere House now houses an angling museum.

Bunbrosna

Mount Murray Isd

Portnashangan

Church Isd

Portlomai

Srudarra Isd

Brown's Isd

Lady's Isd

Royal Canal Feeder

Tullaghan

Mullingar

Lough Owel

(28 kilometres)

LOCATION
Approximately 4 km north of Mullingar off the N4 road

LENGTH
6 km/3.7 miles

WIDTH
3 km/1.9 miles

AREA
1,019 hectares/10.2 km²

PUBLIC ACCESS
Tullaghan, Mullaly's beside the yacht club (both at the south-east end of the lake) and Portnashangan (N4 access)

MAP
Ordnance Survey of Ireland Discovery Series Map 41, covering parts of Longford, Meath and Westmeath (ISBN 978-1-901496-30-7)

Lough Owel has a magical history. Local folklore relates its origins to a dispute between two giantesses who lived in the midlands, on different sides of the River Shannon. The lake was sent on the wind by one sister to the other, on a temporary loan. However, the sister to whom it was sent was so enchanted by its beauty at its new location that she refused to return it, despite many requests.

Profile

The local ballad 'The Three Lakes', referred to earlier describes the lake thus:

> A bright sparkling gem to the North lies:
> Lovely Owel, so serene and so still.
> There Turgesius the Dane
> Met his fate by the lane
> That still winds near the steep Captain's Hill.

The lake is spring-fed and is remarkably clear. On a bright day, it is often possible to see down to a depth of more than six metres. Perhaps because of its water quality, the lake has proved far more useful than its southern counterpart for the people of Mullingar and the surrounding area. In the nineteenth century, it was the scene of an annual festival which took place on the first Sunday of August, known as 'Lough Sunday'. The festival featured swimming races with a difference: the races were performed on horseback. A more useful function for the town has been the supply of drinking water, and there is a large water-treatment plant on the western shore at Portloman. The lake is also the principal water supply for the Royal Canal, via a 4 km feeder channel which runs from a sluice house located beside the sailing club on the southern shore to link up with the main branch of the canal in the town. The feeder is a narrow and relatively shallow channel that is not suitable for navigation by boats. However, a path runs alongside the feeder for most of its course to Mullingar, and this provides an interesting off-road extension to the Lough Owel circuit.

Whereas Lough Ennell is surrounded by relatively flat countryside, Lough Owel appears to sit snugly in the midst of rising ground along its shores, affording the lake surface some degree of shelter – hence the description in the ballad. For this reason, the Lough Owel circuit presents a more varied route, with several ascents and descents, particularly around its northern and western shores. On the circuit, you will encounter several locations where the lake can be admired from a height – none better than the recommended starting point for the circuit at the public parking facility on the N4 at Portnashangan on the south-eastern shore. This is a popular location for swimmers. The lakeshore can be accessed using the footbridge across the railway line which runs along the fringes of the eastern shores of Lough Owel for several kilometres.

Church Island

There are several islands visible from Portnashangan. The largest and closest of these is Church Island, which at one time accommodated a monastery – whose ruins still remain. If the level of the water falls significantly, it is possible to make out a causeway that leads out to the island.

Leaving the Portnashangan car park, turn left on to the N4. The hard shoulder on this busy road affords reasonable comfort and respite from the passing traffic. After about 2 km, you pass Ballinafid Lake, which is sandwiched between the road and the railway line, and then the Covert pub, which lies opposite the lake. About 2 km further on, you have the chance to make a temporary diversion from the busy road by availing of a small country lane to the left. This lane will bring you closer to the lake and provide you with a view across its northern shore. Returning to the N4 for half a kilometre, turn left at Tormey's pub to start tracking the western shore.

There are no public-access points along the lake's western shore. The road that runs parallel to the lake has two turns leading towards the lake. You will encounter the first of these after a steady but gentle climb past Frewin Hill: this leads down to the Portloman water-treatment works. Adjacent are the ruins of an old church and a disused graveyard, where the oldest legible gravestone dates back to 1786. A plaque at the entrance to the graveyard commemorates a pilgrimage and mass held here on 21 May 2000. On the lakeshore, there is also a small private harbour for sailing dinghies. The second turn to the left leads down to two private dwellings, and although there is a castle and motte indicated on the map close to the shore, they are not immediately evident, and it is not possible to inspect the area more closely, as these are private grounds.

The quiet country road on raised ground above the lake leads to the R393, which it meets at Strand crossroads, where you bear left. Less than half a kilometre along is a left turn that will take you, on another quiet road, to a signpost marking the public shore access at Tullaghan. This is a popular area with anglers, with a slipway for launching small fishing boats. Not far along the southern shore, you will be able to see Lough Owel Sailing Club, where there is another public-access facility. Do not be tempted, as I was on one occasion, to try to make your way along the shoreline, as you will quickly get bogged down in marshy ground. The better option is to make your way there by road. The sluice house that stands over the start of the feeder channel for the Royal Canal lies to the left of the car park near the sailing club.

Turgesius

Returning to the road, it is a short ride back over the railway, crossing to the starting point and passing by a hill called Captain's Hill, mentioned in the above ballad as the location where the Danish chief Turgesius is reputed to have been drowned by Malachy I, an Irish chieftain who had captured him on the shores of Lough Ree in 844. There are various legends relating to Turgesius and this event. He is said to have been the most powerful Viking active in Ireland in the early to mid-ninth century. According to one story, he captured Armagh and assumed the position of abbot there, having expelled the previous occupant. He is said to have been in the process of assuming control over the whole of Ireland before Malachy got his hands on him. Donnchadh Ó Corráin, in *Milestones of Irish History*, published in 1986, suggests that these tales are a little far-fetched. Apparently, Turgesius was not as powerful or all-conquering as is suggested, and the story of his conquering ways was in fact political propaganda disseminated by the O'Brien clan in the twelfth century in order to blacken the name of the Uí Neill and to suggest that they had done a bad job in defending Ireland from the Vikings, unlike the O'Briens' own much-heralded ancestor, Brian Boru.

Inny River

Coolure
Demesne

Inny River

Donore

Multyfarnham

Faughalstown

Ringtown

Knockeyon
Mountain

Knockbody
Mountain

Crookedwood

Lough Derravaragh

(45 kilometres)

LOCATION
Approximately 12 km north of Mullingar with the village of Multyfarnham on its southern shore and Castlepollard to the north-east

LENGTH
9.6 km/6 miles

WIDTH
4 km/2.5 miles at its widest point

AREA
1107 hectares/11.0 km^2

PUBLIC ACCESS
Donore, Coolure (closest to Castlepollard), Faughalstown and Gartlandstown (near Crookedwood)

MAP
Ordnance Survey of Ireland Discovery Series Map 41, covering parts of Longford, Meath and Westmeath (ISBN 978-1-901496-30-7)

The name 'Derravaragh' is derived from the Irish word dairbhreach, which translates as 'abundance of oaks' and aptly describes the setting for the third of Mullingar's lakes. The lake is long and narrow, with the northern part broad and shallow and its narrow southern extremity scenically sandwiched between two hills, Knockeyon and Knockbody. The taller of the two hills, Knockeyon, at 215 m, used to be climbed by pilgrims in much the same way as Croagh Patrick is today. The lake used to be larger in area but arterial drainage in the 1960s caused the water level to fall by about six feet. This, combined with a deterioration in water quality and the introduction of roach, has had an impact on the lake's reputation as a trout fishery. However, the drainage has had a positive impact, with an improvement in the lake's ability to act as a reservoir for the River Inny, which flows in and out of the lake's northern end. The lake fills up during winter faster than it drains out, which in turn helps prevent flooding in the surrounding region.

The Children of Lir

The lake is best known for its association with the legend of the Children of Lir, a tale which involved a jealous stepmother changing three children into swans and condemning them to three three-hundred-year periods of enchantment, the first of which was to be spent on Lough Derravaragh. There is a delightful bronze sculpture depicting a scene from the legend on display in the centre of the nearby village of Castlepollard. The legend is briefly inscribed in four languages on a stone plinth in front of the sculpture, which was created by Dolores Nally and was unveiled in April 1999.

Crookedwood

The starting point for the lake circuit is the public-access facility near the village of Crookedwood on the R394 road from Mullingar to Castlepollard. This is signposted down a small winding lane to the left as you descend the hill from Crookedwood. This curious village name is derived from the Irish phrase cnoc an bhodaigh, which translates as 'the hill of the clown' or 'churlish person' and dates back to the legendary tales of Fionn and the Fianna. The shore-access facility is popular with anglers for overnight camping.

The early part of the circuit follows the R394 in the direction of Castlepollard, with lofty Knockeyon and its surrounding rising ground blocking the view of the lake. In keeping with the philosophy of cycling as much as possible on country lanes and as close as possible to the lakeshore, you should take the first turn left for a kilometre-long diversion that will give you a brief glimpse of the lake before you resume on the R394. Leave this road half a kilometre further on at the left turn just past the grounds of Ringtown Hurling Club. Once again the view of the lake is blocked by rising ground, but the first road to the left leads to the shore-access point at Faughalstown, where there are the ruins of a church. Unfortunately, there are no means of continuing the circuit without backtracking on the access lane to this shore. Resuming a northerly course, you will come to a right-angled turn, where there is a metalled lane that continues on straight. While you will get some fine views of the northern shores, there is no lakeside access using this lane, so continue with the principal road. You have a choice of turning left at the next turn or at the later junction. My preference is to take the first left, as it brings you through a delightful tree-shaded lane after you pass by all the dwellings on the left.

Coolure

A short cycle further on is the second access point to the shore, located at Coolure. There are in fact two access points in quick succession at this corner of the lake. My preference is to take the second one, because you get a much better view of the full length of the lake from the northern shore. In addition, you will be closer to the crannóg which lies a little away from the shore and dates back perhaps as far as the seventh century. The crannóg was surveyed by members of the Department of Archaelogy of University College Dublin during the summer of 2004. Dr Rob Sands of UCD, whom I met at the site, advised me that they regarded the crannóg as being a signficant discovery, yielding good initial results. There is also a ringfort close to the shore that you can inspect with ease.

Another reason for opting for the second access to Coolure is that there is a delightful off-road extension to the circuit that you might wish to take. Instead of returning immediately to the road leading to Coole, turn left as if you were going to continue parallel to the lakeshore. This narrow country lane will eventually lead you past bogland up to a small forest that runs alongside the River Inny as it makes its way towards the lake. Unfortunately, there is no way of crossing the river close to the lakeshore. If you decide to take on this extension, do not be tempted to take a short cut through the forest or the bog back to the roadway. I tried to do this on one occasion and it took me over an hour of uncomfortable trekking, while mostly carrying my bike, through the forest and bog to reach the road. It was one of the longest 'short cuts' I have taken. The better option is to return to the road that links the villages of Coole and Multyfarnham. This is no hardship, as you will find a well-surfaced, quiet road that initally skirts Carriskill Bog and crosses the River Inny twice as it enters and leaves the lake.

Multyfarnham

Multyfarnham is dominated by a Franciscan friary, which has maintained a presence here since 1276. The friary survived the suppression of the monasteries, and in the 1950s the friars established an agricultural college, which still operates today. The friary is famous for the life-sized Stations of the Cross to be found in its grounds, which are worth a visit.

Donore

Turning east at Multyfarnham, there is a well-signposted route towards the lakeshore at Donore, where you will find a thriving caravan park, at the entrance of which is a very welcoming coffee shop. This is open from April to October at weekends, and every day during the peak summer months. Leaving Donore, the quiet road that runs parallel to the lake, passing by the rear of Knockbody Hill until it reaches the R394, will return you to the starting point of your journey at Gartlandstown. Unfortunately, the view of the lake eastwards is obscured by high ground, and there is no direct access to the lakeshore between Donore and Gartlandstown.

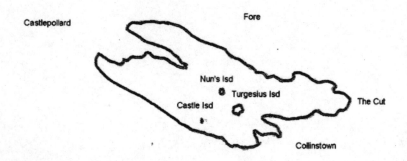

Castlepollard

Fore

Nun's Isd

Turgesius Isd

Castle Isd

The Cut

Collinstown

Lough Lene

(17 kilometres)

LOCATION
Situated between the villages of Castlepollard and Collinstown off the R395, with the historic village of Fore nearby to the north

LENGTH
4.5 km/2.8 miles

WIDTH
1.5 km/1.0 mile at its widest point

AREA
430 hectares/4.3 km^2

PUBLIC ACCESS
'The Cut' near Collinstown at the eastern end

MAP
Ordnance Survey of Ireland Discovery Series Map 41, covering parts of Longford, Meath and Westmeath (ISBN 978-1-901496-30-7), and Map 42, covering parts of Meath and Westmeath (ISBN 978-1-901496-06-2)

In her book Ingenious Ireland, Mary Mulvihill describes Lough Lene as unique, as it drains both west to the Atlantic and east to the Irish Sea. This may sound strange for such a small lake, which lies in the heartland of Ireland away from either coastline. However, she authenticates this claim by revealing that one stream from the lake drains into the River Deel, a tributary of the River Boyne which flows into the Irish Sea at Drogheda. She also states that in the 1980s, geographers from Trinity College Dublin traced an underground channel from the lake to a spring at the village of Fore. This channel drains westward into the River Inny, a tributary of the mighty Shannon.

The lake is part of what has been described as a 'fossil landscape': the remains of an ancient limestone landscape whose hills may once have been rocky limestone towers and pinnacles, and whose lakes are possibly flooded depressions. Its name is an anglicised version of the Irish 'Loch Leibhinn'. Leibhinn was a figure from Irish mythology, being the daughter of the god Manannán Mac Lir and the sister of the sun goddess, Áine. One of the stories told about the lake is that in 864 its waters curiously turned into blood for nine days. What a contrast to the present day, as the lake's waters are now often described as 'gin clear'. The extensive shallows make it a popular lake for swimming, and it has a 'Blue Flag' designation. Lough Lene also has a curious connection with Irish political life. A replica of what is known as the Lough Lene Bell is used to bring the Dáil (the lower house of the Irish parliament) to order.

There is good shore access at a location known as 'The Cut', where there is a car park, a picnic area, toilets and a slipway for boats. From Castlepollard on the R395, bear left at the crossroads at the centre of Collinstown. 'The Cut' is 2 km from the village and is well signposted. It is the optimal starting point for the circuit, particularly as, during the summer months, you can have a refreshing swim on your return.

Fore Abbey

From 'The Cut', make your way north-west along the small country lane which links in with the road to Fore after about 1 km, and bear left. A short cycle ride will bring you to the village itself, which is situated in a valley between two hills. The name of the village is derived from the Irish word fobhar, or 'spring'. The village is dominated by the well-preserved ruins of an ancient monastic settlement which date back to the seventh century and are the largest group of Benedictine remains in Ireland. St Feichin established a monastery here around 630 but the ruins we see today are of later origin. St Feichin is said to have died of the great yellow plague that swept across Ireland in 664. He is also associated with Cong in County Mayo (see page 258). It is surprising that the ruins are in such good condition today, as the abbey had a perilous history. Its earliest community was provided by an abbey in Evreux in Normandy, and because of this French connection Fore was regularly seized by the Crown as alien property when England was at war with France. Between 771 and 1169, Fore Abbey was burnt twelve times.

The Seven Wonders of Fore

The abbey is one of the Seven Wonders of Fore, as it was built in a bog. The other six wonders are the mill without a race, the water that flows uphill, the tree that has three branches (or the tree that won't burn), the water that won't boil, the anchorite (hermit) in a stone, and the stone (or lintel) raised by St Feichin's prayers. The first five of the wonders are located in or around the priory on one side of the road, while the last two are by the ruins of the Church of St Feichin, on the hill on the opposite side. Information on all the wonders is available on information panels at the site, and during the summer months there is a video presentation at the Fore Abbey Coffee Shop.

One of the more impressive and unusual ruins on the Fore Abbey site is the columbarium, or pigeon house, situated on a high position to the north-east. It would have been built by the monks to provide meat over the winter months, and the eggs and droppings (for fertiliser) would also have been useful. The circular structure features lots of small nesting spaces in its walls to accommodate its many residents.

When leaving Fore, keep to the left where the road links up with the R195 to Castlepollard. Once again, the policy of maintaining as close a link to the lakeshore as possible dictates that you take the first turn left off this road, down another delightfully quiet country road. It is possible to access the lakeshore along a small lane leading back from this road. For a time, the lake view is obscured by a small forest, but at the next crossroads turn left to maintain a close connection to the lake. This quiet, undulating road runs parallel to the lake and eventually leads on to the R395 for the short trek to Collinstown. The Irish name for this village is 'Baile na gCailleach', which translates as 'the Town of the Veiled Women'. The description relates to a period during early Christian times when a community of nuns used to live in the area. At the crossroads in the village, turn left and return to 'The Cut'.

County Kerry

The Lakes of Killarney

Lough Leane
Muckross Lake
Upper Lake

It is a daunting prospect to write anything about County Kerry, and in particular Killarney, as many great travel writers and literary giants have already committed their impressions to print. Because of its abundant natural beauty, combining high mountains, dense woodland, meandering rivers, deep lakes and a rocky coastline, the county has been attracting visitors with a passion for the picturesque for several centuries. Many have succumbed to the challenge of describing the awesome scenery, while others have relied on visual representations. Tennyson, Wordsworth, Thackeray and Scott were among the earlier visitors to the county, and all found the scenery inspiring. The mountains, forests and lakes were a natural target for creators of guidebooks, illustrated with topographical engravings, that emerged with the growth of tourism and increasing literacy in the British Isles in the early nineteenth century. Artists like George Petrie (1790–1866) were commissioned to provide drawings from which engravers could produce accurate and picturesque views of buildings and topography for inclusion in these guidebooks. In the early years of the twentieth century, the photographer Louis Anthony was one of the first to produce photographic images of the area, and these were used in a series of postcards that introduced the area's breathtaking scenery to a wider audience. These engravings and photographs enable the modern visitor to compare the present-day topography with that of almost two hundred years ago.

Popularly known as 'The Kingdom', Kerry's name is derived from the Irish 'Ciarraí', which is an abbreviation of the expression 'Ciar Ríacht', translated as 'Ciar's Kingdom'. Ciar was one of triplets born to Queen Maeve of Connacht and her husband Oilill, also known as Fergus Mac Roich. He moved south and created the Kingdom of Kerry. The name 'Ciar' means 'black' and is said to relate to some divine ancestor or totem. The modern county was created in 1606. Previously, south Kerry and west Cork had collectively been known as the County of Desmond.

County Kerry is not all mountainous and rugged. There is an area of undulating plain to the north that extends to the estuary of the River Shannon. Not all comments about the county have been favourable. That great storyteller Frank O'Connor (1903–66) said of the county that its scenery is remarkable 'when you can see it, which, owing to the appalling weather the county enjoys, is very rarely'. Despite the weather, the county's high mountains, rugged Atlantic coastline and lakes continue to appeal to a huge number of visitors every year.

The capital of Kerry is Tralee, but as far as tourists are concerned the principal town is Killarney, which celebrated the 250th anniversary of its founding in 2004. Killarney is replete with history and has hosted many distinguished visitors over the years, including Queen Victoria. It was once described by Thomas Browne, the Fourth Viscount Kenmare (1726–95), as 'a miserable village'. To give him his due, it was he who initiated the development of the place and pioneered it as a tourist destination. However, I would have to agree with Frank O'Connor, who, during his cycles in the area in the 1940s, described it as 'a depressing hole, too wealthy for its lack of taste, and infested by hotel-keepers, touts, jarveys and boatmen, all with a glib flow of patter'. It is simply overtouristed, and this dilutes the enjoyment of some elements of the cycling circuit around its nearby lakes. I am sure that there is a lot more to the town than what is normally presented to tourists, but an attempt to peel off the many layers of tourism can prove too much for the short-term visitor.

River
Laune

Aghadoe

Killarney

Lough
Leane

Knockreer

Brown Isd

Innisfallen

River Flesk

Ross
Castle

Ross Isd

Rough Isd

Tomies
Mountain

Cow Isd

Gap of
Dunloe

Purple
Mountain

Muckross Demesne

Dinish Isd

Muckross
Lake

Meeting of
the Waters

Torc
Waterfall

The Long Range

Torc
Mountain

Owengarrh
River

Upper Lake

Gearhameen
River

Galway's
River

Lough Leane

LENGTH
8 km/5 miles

WIDTH
6.2 km/3.8 miles at its widest point

AREA
1,800 hectares/18 km^2

PUBLIC ACCESS
Ross bay, Tomies East

Muckross Lake

LENGTH
1.5 km/1 mile

WIDTH
4 km/2.5 miles at its widest point

PUBLIC ACCESS
Muckross estate

Upper Lake

LENGTH
0.8 km/0.5 miles

WIDTH
4 km/2.5 miles at its widest point

PUBLIC ACCESS
Lord Brandon's Cottage, plus a number of informal access points on the southern shore

The map covering all three lakes is the Ordnance Survey of Ireland Discovery Series Map 78, covering part of County Kerry (ISBN 978-1-901496-83-3).

The Lakes of Killarney comprise three closely connected lakes that lie towards the eastern end of the Iveragh peninsula in the south of County Kerry. With echoes of similar tales relating to lakes in other parts of the country, local lore relates their origins to the story of a comely maiden who was distracted by the attentions of a handsome young warrior and neglected to replace the cover on a magical well, which was the sole source of drinking water for the local populace. While the people of the valley slept, the unprotected waters of the well gushed forward and flooded the valley, drowning all the inhabitants. The geological explanation is that the lakes were formed by a combination of the cutting effects of glaciers during the Ice Age and the impounding effects of moraines, which are banks of debris built up in the wake of a passing glacier. The largest of the lakes is Lough Leane, also known as the Lower Lake. This is separated by a narrow isthmus from the Middle Lake, which lies within the Muckross demesne, and from which it has drawn its more popular name. A watercourse called 'The Long Range' connects Muckross Lake to the third lake in the chain, the Upper Lake, which has the most dramatic and rugged setting of the three. Situated about 20 m above sea level, the lake waters are drained from the north-west corner of Lough Leane by the River Laune, which enters the sea at Dingle bay near Killorglin.

There is a difference of opinion as to how Lough Leane got its name. According to one local tradition, the lake's name is derived from its association with the island monastery of Innisfallen, which we will come to later. This was a place of learning, and so the lake became known as Loch Léin, which translates from the Irish as 'Lake of Learning'. A more likely explanation for the name is the lake's association with early metalworking, and a mythical goldsmith called 'Lén of the Many Hammers' (also known as 'Lén of the White Teeth'), who served Bóv the Red, king of the Danann of Munster. According to Standish O'Grady in his Critical History of Ireland, published in 1881, Lén plied his trade on the shores of the lake, where he 'wrought, surrounded by rainbows and showers of rainy dew'. Bóv the Red was the king whom Aoife, the second wife of Lir, was going to visit when she transformed the latter's children into swans on the shores of Lough Derravaragh.

The cycling circuit outlined in this section embraces all three lakes in one route. It touches upon sections of both the Ring of Kerry

Cycle Route and the Kerry Way Waymarked Trail, both of which are signposted. The Ring of Kerry Cycle Route starts and finishes in Killarney and extends over 216 km through mountain passes and along Kerry's rugged Atlantic coastline. The Kerry Way also kicks off in Killarney and uses the old droving paths and coach roads of the Iveragh peninsula. A significant part of the circuit in this book avails of routes that lie within Killarney National Park. The park comprises almost 10,300 hectares of woodland, mountains and lakes and is home to a large herd of red deer, Ireland's largest native mammal. The herd within the park is thought to be the purest of any of Ireland's herds of deer. Japanese sika deer are also to be found here. The routes within the park are excellent for cycling and incorporate a rough-surfaced trail along the southern shores of the Upper Lake which is one of the best off-road cycling trails to be found anywhere in Ireland. The circuit also includes a challenging but infinitely rewarding ascent through the Gap of Dunloe and descent into the Black Valley.

I feel it necessary to impart a cautionary word before embarking on the circuit. Cognisant of Frank O'Connor's sentiments in relation to Kerry's climate, you should be aware that the weather can change very quickly around Killarney's lakes. A sunny day can quickly yield to the purifying waters of passing showers, and given that you will be cycling in the midst of Ireland's highest mountains, the elevated sections of the route are susceptible to chilling winds. In addition, during the winter months, when the winter levels of the lakes are high, you may find that some of the paths that run close to the lakeshore are inundated, forcing you to choose an alternative route. Locations that are particularly susceptible to inundation include the pathways near Dinish Cottage on the shore of Muckross Lake, Ross Island on Lough Leane, and the early part of the route of the Kerry Way leading eastward from Lord Brandon's Cottage on the shore of the Upper Lake.

Like a lot of Killarney's tours, I choose St Mary's Cathedral as the starting point for the circuit. Built of limestone in 1845, the cathedral was designed by Pugin. Its distinctive central tower, topped by a spire, serves as a notable landmark for the town and can be seen for miles before the town is reached. Opposite the cathedral is the entrance to the Knockreer section of the national park, which offers a gentle and interesting off-road introduction to the circuit. It can be accessed using the gate beside Deenagh Lodge.

Knockreer demesne

The trails through Knockreer find more favour with the locals than the more celebrated Muckross demesne, and it is easy to understand why. Apart from being quieter, Knockreer has a varied, undulating landscape and offers excellent elevated views over Lough Leane and its most famous island, Innisfallen. A delightful tree-shaded path runs by the banks of the Deenagh river all the way to the lakeshore. Near the lakeshore there is a junction, and you have the option of travelling south towards Ross Castle and Ross Island or crossing the river and continuing north towards Knockreer House using a path described as 'the Circular Walk'. It is intended to keep the visit to Ross Castle to the end of the circuit. A spectacular view of Lough Leane unfolds as you ascend the Circular Walk, with the island of Innisfallen dominating the foreground.

Innisfallen Island

The island's name is an anglicised variant of the Irish 'Inis Faithleann', which translates as 'Faithleann's Island'. Faithleann is thought to have been the son of Aedh Damam, a king of West Munster who reigned in the seventh century. The ruins of a sixth-century abbey are preserved on the island. The abbey is said to have been founded by St Finian, and over the following centuries it grew to become one of the most influential centres of learning in the British Isles. One of its foremost abbots was Maelsuthian O'Carroll, who is described in the Annals of the Four Masters as 'the Chief Saoi (or Sage) of the Western World'. Prior to his death in 1010, he was confessor to Brian Boru, who is believed to have studied on the island before becoming high king of Ireland. The abbot is believed to have been an early contributor to the Annals of Innisfallen, which were compiled on the island and provide a chronicle of Irish history from the earliest times to 1326. The original manuscript of the Annals is housed in the Bodleian Library of Oxford University.

The Sleeping Monk of Innisfallen

One of the legends associated with Innisfallen Island tells the tale of a monk called Brother Cudda, or Cuddy, who was sent on an errand to Muckross Abbey on the mainland. While returning to the island through the grounds of what is now part of the Knockreer estate, he heard a beautiful singing voice and was so mesmerised by the sound that he knelt down to offer a prayer of thanksgiving. He became completely engrossed in listening to the song, and when the singing stopped he emerged from his reverie to discover that the land around him was strangely altered. He could not understand the change. He had left the monastery in the height of summer, yet the landscape had now taken on a wintry profile. Bewildered, he made his way back to the monastery so that he could relate his experiences to his confrères. When he arrived at the gates of the monastery, he was met by an armed stranger who did not recognise him and challenged him when he tried to gain entry. The stranger told him that the monks were no longer in possession of the monastery, adding that Queen Elizabeth had granted the lands to one of her supporters and that it was a dangerous time for men of the cloth. Brother Cudda identified himself to the stranger and insisted that he had only left the island the day before. The stranger denied knowing him but told him of a monk who had gone missing over two hundred years before and was presumed to have drowned in the lake. It was then that Brother Cudda recalled the beautiful singing, and it dawned on him that he had participated in some kind of a miracle.

In one version of the tale, Brother Cudda subsequently made his way to Spain, where he spent his last years. Another version relates that he was shriven and turned to dust as soon as the realisation of the missing years dawned on him. Located beside a holy well in a small wooded area about 200 m off the Circular Walk, within the grounds of the Knockreer Estate, is a large stone with two circular indentations. This is known as 'Cloch na Cudda' or 'Cuddy's Stone', and is said to be where Brother Cudda knelt in prayer while listening to the beautiful singing.

Before you ascend to Knockreer House via the Circular Walk, it is worthwhile making another small diversion at the same point where you left the path to visit Cuddy's Stone. On this occasion, you will need to go in the opposite direction. A rough path leads towards an eminent position known as Cnoc an Rí – the King's Hill – from which the estate derives its name. From this vantage point, there is a fine panoramic view of almost the entire town of Killarney.

Knockreer House

The present Knockreer House is an architecturally bland edifice that does no justice to its magnificent location. It was built by Lady Grosvenor in the 1950s to replace an earlier building which had been destroyed by fire. Red-bricked remnants of the earlier building can be seen close to the present house. The house is now used as an education centre, and during the summer months there is access to an exhibition of the flora and fauna of the locality, and also of the history of the Kenmare family, who occupied the estate in earlier times. From the tiered gardens to the rear of the house there are magnificent views of Lough Leane and south to Ross Castle and the mighty Mangerton Mountain.

Gap of Dunloe

To leave Knockreer, you will need to return to the Circular Walk and follow the rough path that you took to access Cnoc an Rí along the border of the estate, where you can either take the route of the Fossa Way, which leads on to the Killorglin Road (N72), or alternatively skirt past the golf course and join the road further along. You have now joined the course of the Ring of Kerry Cycle Route, which you will need to follow for a couple of kilometres until you turn off for the Gap of Dunloe. Continue to follow this route as it leaves the N72 and heads south-west, with the northern shore of Lough Leane on the left. After crossing the River Laune, bear sharply right at the next junction until you come to another junction opposite the entrance to Dunloe Castle Hotel. Turn left here to start the ascent towards the Gap of Dunloe.

One of the highlights of this circuit is the 9 km ascent through the magnificent glacial legacy called 'the Gap of Dunloe'. The Gap is a 500 m-deep cut carved out between the formidable Macgillycuddy Reeks Mountains to the west and Tomies Mountain and Purple Mountain to the east. The River Loe cascades down through the Gap, gathering the foaming flows of the myriad of streams that run off the surrounding slopes and taking breathers along the way in a chain of small lakes before emptying into the River Laune. The massive boulders that lie strewn across the rough landscape, and the gnarled yet colourful towering slopes, combine to generate an impressive valley with an untamed quality about it.

The Gap is named after Lugaid, son of Cu Roi Mac Daire, one-time ruler of Munster. Also known as 'Lugh', he occupies an important place in Irish mythology in that it was he who hurled the magic spear that killed Cu Chulainn near Dundalk, County Louth, in order to avenge the death of his father. This act is commemorated in a magnificent sculpture in the General Post Office on Dublin's O'Connell Street.

I would have to disagree strongly with the sentiments expressed in 1906 by Baddeley and Ward, who in their guide stated that 'cyclists will not find the Gap pleasant'. I have no doubt that the limitations of early-twentieth-century bicycles and the quality of the road surface at the time had a lot to do with their views. In my opinion, cycling through the Gap is an exhilirating experience, notwithstanding the energy-sapping gradient. The mesmerising scenery will take your mind off the increased effort required, and if all comes to all you can dismount and walk the steeper sections. Indeed, this will offer you the chance to gaze back through the 'V' of the valley, and you may be lucky enough to witness a contrasting scene of the sun shining down on the distant plains to the north.

If you journey through the Gap during the busy tourist season, you may be faced with a frenetic thoroughfare thronged with walkers, ponies, jaunting cars and fellow cyclists, as well as cars. In my view, you will see the Gap at its best during the quieter winter months, when you will be able to savour its wild and wonderful scenery not quite in solitude but in a less crowded atmosphere.

Kate Kearney's Cottage

You should not embark on the ascent through the Gap without first visiting Kate Kearney's Cottage, which jealously guards its entrance. According to local legend, Kate Kearney was a local beauty who used to sell a particularly fiery brand of poteen that she brewed herself in the middle of the nineteenth century. Her brew was described as 'very fierce and wild, requiring not less than seven times its own quantity of water to tame and subdue it'. The practice of distilling and selling poteen was illegal, but this fact did not prevent her cottage from gaining a reputation for providing sustenance to weary travellers, a tradition that has been maintained today in the form of a family-run restaurant and bar that is open all year round.

The Serpent Lake

As you ascend through the Gap, you will pass by a number of small lakes. The first of these, situated just before you come to Echo Bridge, is listed on the map as 'Coosaun Lake' but is better known by the locals as 'The Serpent Lake'. According to local storyteller Donal O'Cahill, St Patrick cast into the lake an iron chest that contained the last serpent in Ireland. St Patrick had to trick the serpent to get it into the chest, and locals will tell you that the forlorn cries of the captive serpent can be heard coming from the depths. They also say that the wind never disturbs the surface of this lake and that the ripples you see on the surface are caused by the serpent's struggles to get free.

As you approach the Head of the Gap through a series of hairpin bends, be prepared for strong gusts of wind, from which you will have been sheltered during your ascent. The descent into the Black Valley is brief but invigorating. Great care should be taken because of the variable quality of the potholed road surface and the sharp hairpin bends, whose cambers are not always favourable. In any event, speed should not be a priority. It is far more rewarding to amble down with a moderate application of brake pressure, so that you will have a chance to savour the unfolding beauty of the Black Valley. The valley takes its name from the darkness caused by the deep shadows cast over it by the mountains of the Macgillycuddy Reeks and the dark peaty brown of its soil. The overall effect is added to by the darkness of the waters of the Gearhameen river.

Lord Brandon's Cottage

At the first junction, bear left and follow the signposts for Lord Brandon's Cottage. The road follows the course of the Gearhameen river through a picturesque setting until you reach an arched gateway, through which you can access Lord Brandon's Cottage. Very little remains of the cottage itself other than a tower and some smaller ruins, but a modern building houses a café, open only during the summer months. For those who do not wish to continue by bicycle, there is a waterborne option for the return journey to Killarney. During the summer months, a boat leaves the dock at Lord Brandon's Cottage at 2.15 PM and takes you through the three lakes to Ross Castle. As the boat has only limited accommodation for bicycles, it is important to book your ticket in advance. This can be done by contacting O'Connors Traditional Pub in High Street, Killarney.

The Upper Lake

If you were to return by boat, you would be missing out on one of the best off-road trails I have encountered on the shores of any lake in Ireland. The trail follows the route of the Kerry Way and maintains close contact with the shores of the Upper Lake for about 4 km. A good deal of the path has been dressed with a rough stone surface, which is ideal for use by hybrid or trail bikes. This route is important from a cyclist's viewpoint as it avoids the necessity of a lengthy 16 km detour via Moll's Gap – at least half of which would have to be undertaken on the busy Killarney–Kenmare road (N71). The course is challenging, and quite apart from operating as a short cut, it allows you to admire at close quarters the rugged beauty of the Upper Lake, which is dotted with islands and surrounded by sharply rising slopes. After initially tracking across level but marshy ground, the trail adopts an undulating tree-lined course through Derrycunnihy Wood and crosses several small streams before it reaches Galway's river, where it turns south in order to track up to the N71. For those journeying in a clockwise circuit, it is important to note that there is no signage off the N71 to indicate the existence of this trail at the point where it joins the road. The only indication is a small rough car park, whose entrance is squeezed between two mature trees.

Once you have crossed Galway's Bridge, the winding N71 will take you back to the shores of the eastern portion of the Upper Lake, where there are excellent views down the full length of the lake. The scenery continues to amaze, with an abundance of colour even on the greyest of winter days. At the lakeshore near Stag Island, you pass under a crudely carved archway. Along this road, you are also likely to encounter wandering sheep grazing the long acre. It always amazes me how they are completely oblivious to the dangers of speeding cars yet when a cyclist comes along they scarper in mild panic!

Torc waterfall

As stated earlier, the Upper Lake and the Middle or Muckross Lake are connected by a watercourse, in the midst of which is an expanded section known as 'the Long Range'. You will only be able to catch glimpses of this river until you reach Five Mile Bridge at the northern end of the Long Range. From here, the road turns gently leftward as you approach Muckross Lake. There is a small car par park off the N71, from where you can access a trail that leads around the lake. However, before embarking on this trail, a visit to Torc waterfall is recommended. This is located 400 m off the N71, approximately 1.5 km further east, and is well signposted. The winter months are the best time to visit: the Owengarrh river is at its fullest and can provide the waterfall with a supply of water that will do justice to its reputation as one of Ireland's finest major falls. There is a viewing point up a steep path, where you can watch the foaming white water cascading down the black rock face.

The meeting of the waters

There is an entrance to Muckross demesne opposite Torc waterfall. However, in order to complete the circuit around Muckross Lake, it is necessary to backtrack to the entrance passed earlier on the N71. There is a well-maintained, narrow, car-free roadway running by the lakeshore past Bog bay to Dinish Cottage. A path leads behind this cottage to a scenic area known as 'the Meeting of the Waters', from where you can see another famous landmark, the Old Weir Bridge, which crosses the rapids of the Long Range river. The road continues through Dinish Island towards Brickeen Bridge, which spans another link between Muckross Lake and Lough Leane. During the winter months, particularly after periods of heavy rainfall, this road can become quite badly inundated, preventing any chance of passage, even for the most adventurous of cyclists – or, indeed, walkers. On one occasion when I attempted to use the road, the level of the water in a dip not far from Dinish Cottage was almost waist-high and had completely covered one of the signposts.

Muckross House, gardens and traditional farms

There are excellent views of both lakes as you continue east deeper into Muckross demesne, skirting Doo Lough and passing through one of the largest pure yew woods to be found in Europe, near the shores of Kilbeg and Dunday bays. Yew trees can survive for many hundreds of years and are regarded as the oldest living things in Europe. They have always been popular in Ireland and were regularly planted in sacred places and cemeteries. The association with graveyards comes from the long-held belief that it was only safe to plant them near the dead because of their poisonous leaves. Mary Mulvihill, author of Ingenious Ireland, believes that a yew tree reputedly planted in 1344 at Muckross Abbey (which we will come to soon) is the oldest living thing in Ireland. The trail leads to Muckross House, which was built in 1843 by the Herbert family and today is preserved as a period home. (The Herberts were among a number of English planters who were granted lands in the Kerry region at the end of the sixteenth century.)

The house was once owned by Arthur Guinness, who also owned Ashford Castle on the shores of Lough Corrib. He bought it to prevent it from falling into the hands of a syndicate of English property developers. It is open daily all year round and is subject to an entrance charge. A restaurant and craft centre are incorporated into the house. The gardens beside the house are extensive, and there is free public access. The traditional farms, comprising three separate farms of varying sizes, cover 30 hectares and display traditional farming methods of an earlier age. There is an entrance charge to the farms.

Muckross Abbey

Returning to the trail, Muckross Abbey can be found to your right as you make your way to the eastern exit of Muckross demesne. Legends abound in relation to the abbey and its founder, Donal Mc Carthy Mór. According to the Annals of the Four Masters, the abbey, whose correct name is 'the Abbey of Irrelagh' (sometimes recorded as 'Irrilagh'), was founded in 1340 by Mc Carthy Mór for the Franciscan friars, but it is more widely accepted that it was built over a century later, between 1448 and 1475. It is located at a place called Carraig-an-Cheoil ('the Rock of Music'), and local legend refers to the abbey having been built on a song. The legend is told in a lengthy poem composed by Reverend Arthur Blennerhasset Rowan (1800–61), who describes an ageing and penitent Mc Carthy Mór seeking forgiveness for wrongdoings in his fighting past. The elderly chieftain had consulted a holy friar about how he might make amends for all his misdeeds and was commanded to build a church where he should do penance and where prayers could be offered for those he had killed. When he enquired as to where he might build the church, he was told to go eastward:

Seeking the rock by fate decreed,
Where fane may rise and priests can pray.

Dressed as a pilgrim, Mc Carthy Mór wandered wearily about his kingdom seeking the elusive rock, until one summer's day he arrived at a sheltered bay on the southern shores of Lough Leane, described in the poem as 'a diamond set in a ring of rock and wood'. While he was resting at this spot, he saw a young girl approaching a nearby well, pitcher in hand and singing a sweet song, unaware that she was being watched. The song touched the chieftain deeply and awoke in him distant memories of his own childhood, when he had been able to roam the woods carefree and innocent. Spellbound by the haunting music, he approached the young girl and noticed that she was sitting on a rock beside a well. Immediately, he knew that his search had ended, for he had found the 'rock of music', and it was here that he would build his church.

The abbey flourished for about a century before being suppressed in 1541. In 1587, Queen Elizabeth I granted a lease of the abbey to a descendant of the founder also called Donal, who, in addition to the Irish title of 'the Mc Carthy Mór', now also held the title of 'Earl of Clancar' (sometimes recorded as 'Clancarty'). Despite the terms of the lease decreeing that the abbey should be put to secular use, he allowed the monks to continue to live at the abbey. In 1595, the abbey and its lands were granted to a Captain Collum, and it is probable that at this time it was held as an English fortress. In the early years of the seventeenth century, the monks returned to the abbey, and efforts were made to restore it to its original splendour. Except for brief periods, they remained in occupation until Cromwellian soldiers finally drove them out in 1652. There are reports of a small community returning to the abbey for a while in the late eighteenth century.

The abbey ruins are very well preserved and are in better condition now than is depicted in the engraving from an 1822 sketch by George Petrie contained in the Reverend G. N. Wright's A Guide to the Lakes of Killarney, published in 1823. In that engraving, the walls of the interior are depicted as being covered with a large amount of foliage. One of the most impressive areas of the abbey is the cloister, surrounded by an arcade, which is different on each side, probably (given its patchy history of occupation and restoration) due to different phases of construction. An old yew tree dominates the centre of the cloister.

McCarthy Mór's Castle

After you leave the grounds of Muckross demesne, there is a tarma-cadam lane that runs parallel to the N71 and leads most of the way back to Killarney. During the high tourist season, this lane is used by the jaunting cars, so care is needed. As you head towards the town, there are only intermittent views of the Castlelough bay area of Lough Leane to your left. Lying within the grounds of the Lake Hotel are the ruins of a medieval castle, known as 'Castle Lough' – or 'Castlelough', as it has now become – which was once one of three castles owned by the McCarthy clan in south Kerry. The castle occupies an excellent defensive position on a small promontory that reaches out from the lakeshore. In the nineteenth century, a fine country house was built nearby. This house was later converted into a hotel and passed through various hands before being purchased in 1940 by the Huggard family of Waterville, who continue to operate it as an hotel. Permission to visit the castle ruins should be obtained from the Lake Hotel.

Ross Castle

Just before you reach Killarney, turn left off the N71 on to Ross Road in order to visit one of the most historic and impressive edifices to be found on the shores of Lough Leane. Ross Castle is magnificently sited at the water's edge, in a sheltered bay that is protected from the south by Ross Island and to the north by the outer extremities of Knockreer demesne. It is probable that a fortification of some description has been on this site since Norman times, but the struc-ture which has survived is mostly associated with the O'Donoghues, who ruled the Killarney area, holding their lands from Mc Carthy Mór. Over the years, there have been many modifications to the castle, as it changed both hands and purpose, but its striking Norman-style keep has been a constant feature. The tower is said to have been built in the sixteenth century but there is continuing debate as to who was respon-sible for it. Local tradition suggests the O'Donoghues but some believe that it was built by others, possibly the Fitzgeralds, one of whose number was created Earl of Desmond by Edward II in 1329, and that the O'Donoghues then seized it when they held the upper hand in the area.

Over the years, Ross Castle changed hands several times. The O'Donoghues forfeited the castle and neighbouring lands for rebelling against the English, and it was claimed by the McCarthys. They in turn leased it to the Browne family, who were subsequently awarded possession of it in 1588 by way of royal grant after the death of the Mc Carthy Mór. The Brownes resided in the castle up to the middle of the seventeenth century, when the castle became the centrepiece of the final struggles of the Cromwellian wars of 1641–52. At the time, the property was owned by the Third Earl of Kenmare, Valentine Browne, who was a minor. The Brownes had sided with the Catholic Irish forces during the wars, and Ross Castle was the place to which Donogh McCarthy (Lord Muskerry) headed for a last stand after he had been defeated at Knocknaclashy in July 1652. He was pursued by General Ludlow, who had intended to lay siege to the castle. Due to its location, the castle was easy to defend from the land side, but it is said that it fell without a siege due to superstition. The besieged soldiers were aware of an old prophecy that said that Ross Castle would never be taken until strange ships appeared on the Lakes of Killarney. Ludlow was aware of the strength of the castle on its landward side and so arranged for a number of ships to be hauled from the mouth of the River Laune to Lough Leane. These ships were then filled with troops and launched into the lake, heading for Ross Castle. On seeing the ships, the besieged soldiers surrendered without a struggle.

The castle and lands were restored to the Brownes by Charles II in the 1660s but they were to lose possession once again, for siding with James II against William of Orange during the Williamite campaign of 1689–91. They bought them back in the early years of the eighteenth century and in 1721 built a new residence called 'Killarney House' on an elevated site very close to the junction between the Muckross and Ross roads. This house was demolished in 1870 and a more grandiose structure was erected. Unfortunately, this was later destroyed by fire and only a few remnants remain.

Ross Castle is very well preserved and is open to the public during the peak tourist season, subject to an entrance charge. Motor-boat trips can be taken from the quayside beside the castle to Inisfallen Island, and to the Meeting of the Waters and Muckross House and Gardens.

O'Donoghue Ross

Many legends exist in relation to a former resident of Ross Castle who is now known in folklore as 'the Chieftain of the Lakes'. It is said that when he was enchanted, he jumped from one of the windows of the castle into the waters of the lake and that he now presides over an underwater palace at the bottom of the lake. The story behind this leap is one of the most-repeated pieces of folklore in the region. O'Donoghue was a powerful man in every respect and was known to dabble in the black arts. He had one failing, however: a fear of growing old. As the years gathered, he spent more and more time searching in vain for an elixir of youth. Eventually, he locked himself away with his books and potions in a room at the top of the castle. He toiled away for seven weeks before he emerged, shouting for his wife to join him. When his wife arrived at the top of the castle, he told her of his research and plans, and that he was now ready to conduct a number of tests but that he needed her help. He asked her to join him in his room and showed her a large cauldron. He added that he needed her to cut him up into small pieces, put them into the cauldron, and then leave the room, firmly bolting the door behind her, and not return for seven weeks. He promised her that when she returned he would be alive and well and thoroughly rejuvenated. While his wife agreed to his requests, he felt it prudent that he should first put her through a number of tests, as he warned her that should she fail to complete the main task, there would be barely anything left of him. He told her that he was going to read to her from a large black leather-bound tome, and he warned her that should she cry out at anything she heard or saw, he would be taken away from her, never to be seen again. While he was reading, the room was filled with horrific images and sounds, but his wife showed no reaction. Suddenly, the image of her own child lying dead on a table was conjured up before her. She could not contain her grief and let out a terrible wailing cry, at which the room shook heavily and O'Donoghue was seen to leap out of the window, disappearing into the lake below. At the same time as he jumped, his white horse, his table and his library also disappeared. It is said that these turned to stone and can be seen at various points around the lake. Near the western edge of Ross bay is a rock known to this day as 'O'Donoghue's Library'.

O'Donoghue is said to emerge from his watery palace on his white horse on the first of May every seventh year in order to check on his lands. Those who are lucky enough to encounter him are said to be blessed with good fortune thereafter.

Ross Island

South of Ross Castle is Ross Island, a sixty-hectare promontory that contains the site of a 4,500-year-old copper mine said to be the oldest in north-west Europe, and also the first site where metal was made in the British Isles. The mines are located on the southern shore of the island, looking across towards Muckross Lake, and can be accessed via a tarmac track that also leads to a delightful viewing point called Governor's Rock, to the west of the island. In winter months, this track can become badly inundated, preventing access to these points. The mines were operated commercially for about a quarter of a century at the beginning of the nineteenth century and as many as five hundred miners were employed on the site. Underground tunnels were carved out beneath the lake level, and these were accessed via vertical shafts. If you visit in winter, it is easy to understand how flooding became a persistent problem, and at one stage there was a proposal to drain Lough Leane in order to dry the mines. Fortunately, an alternative method of damming the flow of water into the shafts was found, and a large drainage pump was installed. Evidence of the flooded shafts and the dams can still be seen as you progress around the self-guided Mining Trail.

The miners in the eighteenth and nineteenth centuries discovered evidence of earlier works which were widely thought at the time to have been associated with the Danes. However, in 1992 Galway University established a project to investigate the history and archaeology of the Ross Island site, and through the use of radiocarbon date testing it has been established that the early phase of mining there took place between 2400 and 1800 BC. This makes Ross Island the oldest copper mine presently known in western Europe. It was further established that the copper-smelting furnaces and associated slag deposits date back to the seventh and eighth centuries AD. This is the first recorded evidence of primary copper production dating from the early Christian period.

Close to Governor's Rock there is a rough path leading to the rock known as 'O'Donoghue's Library', mentioned above, from where there is an excellent view of Innisfallen Island.

There is a very pleasant lakeshore route from Ross Castle back into Knockreer demesne, where you can make your way back to St Mary's Cathedral.

Select Bibliography

Annals of Loch Cé
Annals of the Four Masters
Annals of Innisfallen
Barrington, T. J., Discovering Kerry: Its History, Heritage and Topography (Cork, 1999)
Barrow, Lennox, Irish Round Towers (Irish Heritage Series No. 8) (Dublin, 1976)
Bulfin, William, Rambles in Eirinn (Dublin, 1907)
Byrne, Patrick, Irish Ghost Stories (Dublin, 1965)
Carra Historical Society, The Moores of Moore Hall (Castlebar, 1989)
Coillte, Coillte Recreation Policy Document (Wicklow, 2004)
Columb, Frank, Lough Gowna Valley (Oxford, 2002)
Conwell, John Joseph, A Galway Landlord During the Famine: Ulick John de Burgh, First Marquis of Clanricarde (Dublin, 2003)
Corcoran, Mary & Others, Monasteraden: Its Past and Its People (Monasteraden, 2004)
Crofton Croker, Thomas, Fairy Legends and Traditions of the South of Ireland (London, 1828)
Crofton Croker, T. and Sigerson Clifford, Legends of Kerry, Kerry, 1975
Cumberlidge, Jane, The Inland Waterways of Ireland (Cambridgeshire, 2002)
Egan PP, Rev Thomas A, The Story of Ballintubber Abbey (Mayo, 2001)
Ferguson, Andrew, Lough Melvin: A Unique Fish Community (Belfast, 1985)
Fredengren, Christina, Crannogs (Wicklow, 2002)
Glover, Winifred, Exploring the Spanish Armada (Dublin, 2000)
Harbison, Dr Peter, Guide to National and Historic Monuments of Ireland, Dublin, 1992
Madden, Gerard, Holy Island: Jewel of the Lough (Clare, 1990)
Massey, Eithne, Legendary Ireland (Dublin, 2003)
Moriarty, Christopher, Irish Lakes (No. 70 of the Irish Environmental Library series) (Dublin, 1988)
Mulvihill, Mary, Ingenious Ireland (Dublin, 2002)
Murray, Peter, George Petrie (1790–1866): The Rediscovery of Ireland's Past, (Cork, 2004)
Ó Corráin, Donnchadh, Milestones in Irish History (Dublin, 1986)
O'Farrell, Padraic, Irish Ghost Stories (Dublin, 2004)
O'Sullivan, Aidan, The Archaeology of Lake Settlement in Ireland (Discovery Programme Monographs 4)
O'Sullivan, Aidan, Crannogs: Lake-dwellings of Early Ireland (Irish Treasures series) (Dublin, 2000)
Rogers, Mary, Prospect of Erne (Fermanagh, 1967)
Rolleston, T. W., Myths & Legends of the Celtic Race (London, 2004)
Rothery, Sean, A Field Guide to the Buildings of Ireland (Dublin, 1997)
Sweetman, David, The Medieval Castles of Ireland (Cork, 1999)
Waldron, Jarlath, Maamtrasna: The Murders and The Mystery (Dublin, 1992)
Warner, Dick and Niall Fallon Waterways: By Steam Launch Through Ireland (London, 1995)
Wilde, Sir William R., Lough Corrib: Its Shores and Islands (Galway, 2002)
Wilkinson, Sidney Berdroe, Reminiscences of Sport in Ireland (London, 1931)
Wilson, Derek, Dark and Light: The Story of the Guinness Family (London, 1998)

Index

Win a different kind of holiday!

Liberties Press & Emerald Star offer you the chance to win a luxury holiday!
ESCAPE TO THE SHANNON!

Boating on the River Shannon offers the perfect antidote away from the hustle and bustle of modern Ireland. A cruise on the River Shannon allows for a truly peaceful, tranquil and relaxing break – the ideal way to 'chill out' with family or a group of friends. Just imagine being the master of your own boat for a full week – no hassles, no worries – hopping from village to village!

Emerald Star, Ireland's Largest Inland Waterway Cruiser Company and Liberties Press are delighted to offer you a chance to win a week's holiday on the Caprice - an amazing luxury cruiser for four people.

Once on board you will be transferred into a world of total tranquillity and luxury. This beautifully furnished cruiser is designed for those who only insist on the best!

Question : How many bases do Emerald Star have in Ireland?

Answer: _____

Name: _____

Contact number: _____

Address: _____

E-mail: _____

For more information on Emerald Star's fabulous boating holidays, log on to www.emeraldstar.ie. Send your entry to Emerald Star, The Marina, Carrick on Shannon, County Leitrim, Ireland

Terms and conditions :

Subject to availability. The first correct entry drawn will be the winner. Competition closing date will be 01 September 2008. Employees and families of Liberties Press and Emerald Star are not eligible to enter. Winner must be 18 years or over and agree to Emerald Star conditions of hire, including damage waiver insurance OR refundable damage waiver insurance & security deposit. No cash in lieu of prize. Departure and return base is Portumna.

Holiday prize is valid for travel from 1 September 2008 to 31 May 2009, excluding bank holiday weekends. Diesel not included in the prize. Winner to be notified in writing,